An Introduction to Decision Theory

This up-to-date introduction to decision theory offers comprehensive and accessible discussions of decision making under ignorance and risk, the foundations of utility theory, the debate over subjective and objective probability, Bayesianism, causal decision theory, game theory and social choice theory. No mathematical skills are assumed, and all concepts and results are explained in non-technical and intuitive as well as more formal ways. There are over 100 exercises with solutions, and a glossary of key terms and concepts. An emphasis on foundational aspects of normative decision theory (rather than descriptive decision theory) makes the book particularly useful for philosophy students, but it will appeal to readers in a range of disciplines including economics, psychology, political science and computer science.

MARTIN PETERSON is Associate Professor of Philosophy at the Royal Institute of Technology, Sweden. He is author of *Non-Bayesian Decision Theory* (2008).

An Introduction to Decision Theory

MARTIN PETERSON

Royal Institute of Technology, Sweden

CAMBRIDGE
UNIVERSITY PRESS

CAMBRIDGE UNIVERSITY PRESS
Cambridge, New York, Melbourne, Madrid, Cape Town, Singapore, São Paulo,
Delhi, Dubai, Tokyo, Mexico City

Cambridge University Press
The Edinburgh Building, Cambridge CB2 8RU, UK

Published in the United States of America by Cambridge University Press, New York

www.cambridge.org
Information on this title: www.cambridge.org/9780521716543

First published 2009
Reprinted 2010

Printed in the United Kingdom at the University Press, Cambridge

A catalogue record for this publication is available from the British Library

ISBN 978-0-521-88837-0 hardback
ISBN 978-0-521-71654-3 paperback

Contents

Preface

This book is an introduction to decision theory. My ambition is to present the subject in a way that is accessible to readers with a background in a wide range of disciplines, such as philosophy, economics, psychology, political science and computer science. That said, I am myself a philosopher, so it is hardly surprising that I have chosen to discuss philosophical and foundational aspects of decision theory in some detail. In my experience, readers interested in specific applications of the subject may find it helpful to start with a thorough discussion of the basic principles before moving on to their chosen field of specialisation.

My ambition is to explain everything in a way that is accessible to everyone, including readers with limited knowledge of mathematics. I therefore do my best to emphasise the intuitive ideas underlying the technical concepts and results before I state them in a more formal vocabulary. This means that some points are made twice, first in a non-technical manner and thereafter in more rigorous ways. I think it is important that students of decision theory learn quite a bit about the technical results of the subject, but most of those results can no doubt be explained much better than what is usually offered in textbooks. I have tried to include only theorems and proofs that are absolutely essential, and I have made an effort to prove the theorems in ways I believe are accessible for beginners. In my experience, this sometimes comes into conflict with the ambition to present technical material in the minimalist style usually preferred by experts.

Most of the technical results are presented in twenty 'boxes' clearly separated from the main body of the text. In principle, it should be possible to read the book without reading the boxes, although they hopefully deepen the student's understanding of the subject. I have also included over 100 exercises (and solutions), most of which should be fairly straightforward. Unlike other textbooks, no exercise asks the reader to prove some theorem

I did not bother to prove myself. Finally, Appendix A contains a glossary in which I try to briefly explain some of the key terms and concepts. I believe the glossary might be particularly useful for readers wishing to study only a selected number of the chapters.

A large number of people deserve my sincere thanks. First of all, I would like to thank all students who have contributed their invaluable input to this project. I have done my best to improve the manuscript in accordance with the advice I have received. I am also deeply indebted to a number of fellow teachers and colleagues: Barbro Björkman, Joanna Burch Brown, John Cantwell, Stephen John, Elselijn Kingma, Holger Rosencrantz, and Per Sandin. I am also very grateful for valuable comments on the original proposal and draft manuscript given by four anonymous readers. Finally, I wish to thank Hilary Gaskin at Cambridge University Press, who suggested I should write this textbook. Without her enthusiasm and encouragement, this book would never have been written. The project proved to be both challenging and time consuming but always highly enjoyable.

1 Introduction

On 6 September 1492 Christopher Columbus set off from the Canary Islands and sailed westward in an attempt to find a new trade route between Europe and the Far East. On 12 October, after five weeks of sailing across the Atlantic, land was sighted. Columbus had never been to the Far East, so when he landed in Middle America ('the West Indies') he believed that he had indeed discovered a new route to the Far East. Not until twenty-nine years later did Magellan finally discover the westward route to the Far East by sailing south around South America.

Columbus' decision to sail west from the Canary Islands was arguably one of the bravest decisions ever made by an explorer. But was it rational? Unlike some of his contemporaries, Columbus believed that the Earth is a rather small sphere. Based on his geographical assumptions, he estimated the distance from Europe to East India to total 2,300 miles. The actual distance is about 12,200 miles, which is more than five times farther than Columbus thought. In the fifteenth century no ship would have been able to carry provisions for such a long journey. Had America not existed, or had the Earth been flat, Columbus would certainly have faced a painful death. Was it really worth risking everything for the sake of finding a new trade route?

This book is about decision theory. Decision theory is the theory of rational decision making. Columbus' decision to set off westwards across an unknown ocean serves as a fascinating illustration of what decision theory is all about. A *decision maker*, in this case Columbus, chooses an act from *a set of alternatives*, such as sailing westwards or staying at home. The *outcome* depends on the true *state of the world*, which in many cases is only partially known to the decision maker. For example, had the Earth been a modest-sized sphere mostly covered by land and a relatively small and navigable sea, Columbus' decision to sail westwards would have made him rich and famous, because the King and Queen of Spain had promised him ten per cent of all revenue gained from a new trade route. However, Columbus' geographical hypothesis turned out to be false. Although spherical, the Earth is much bigger than Columbus assumed, and Europe is separated from the Far East by a huge continent called America. Thus, in the fifteenth century the westward route was not a viable option for Europeans wishing to trade with the Far East. All this was unknown to Columbus. Despite this, the actual outcome of Columbus' decision was surprisingly good. When he returned to Spain he gained instant fame (though no financial reward). Another possible outcome would have been to never reach land again. Indeed, a terrible way to die!

The decision problem faced by Columbus on the Canary Islands in September 1492 can be summarised in the *decision matrix* shown in Table 1.1. Note that the outcome of staying at home would have been the same no matter whether his geographical hypothesis was true or not.

Since the second hypothesis turned out to be the true one, the actual outcome of sailing westwards was that Columbus got famous but not rich. However, it should be evident that the rationality of his decision depended on *all possible* outcomes – all entries in the matrix matter. But how should one use this basic insight for formulating more precise and useful theories about rational decision making? In this book we shall consider a number of influential attempts to answer this question.

Roughly put, the ultimate aim of decision theory is to formulate hypotheses about rational decision making that are as accurate and precise as possible. If you wish to tell whether Columbus' decision to sail westwards was rational, or whether this is the right time to invest in the stock market, or whether the benefit of exceeding the speed limit outweighs the risk of getting caught, then this is the right subject for you. You say it is not worth the effort? Well, that is *also* a decision. So if you wish to find out whether

Table 1.1

	Geographical hypothesis true	There is some other land westwards	There is no land westwards
Sail westwards	Rich and famous	Famous but not rich	Dead
Do not	Status quo	Status quo	Status quo

the decision not to learn decision theory is rational, you must nevertheless continue reading this book. Don't stop now!

1.1 Normative and descriptive decision theory

Decision theory is an interdisciplinary project to which philosophers, economists, psychologists, computer scientists and statisticians contribute their expertise. However, decision theorists from all disciplines share a number of basic concepts and distinctions. To start with, everyone agrees that it makes sense to distinguish between *descriptive* and *normative* decision theory. Descriptive decision theories seek to explain and predict how people *actually* make decisions. This is an empirical discipline, stemming from experimental psychology. Normative theories seek to yield prescriptions about what decision makers are *rationally required* – or *ought* – to do. Descriptive and normative decision theory are, thus, two separate fields of inquiry, which may be studied independently of each other. For example, from a normative point of view it seems interesting to question whether people visiting casinos in Las Vegas *ought* to gamble as much as they do. In addition, no matter whether this behaviour is rational or not, it seems worthwhile to *explain* why people gamble (even though they know they will almost certainly lose money in the long run).

The focus of this book is normative decision theory. There are two reasons for this. First, normative decision theory is of significant philosophical interest. Anyone wishing to know what makes a rational decision rational should study normative decision theory. How people actually behave is likely to change over time and across cultures, but a sufficiently general normative theory can be expected to withstand time and cultural differences.

The second reason for focusing on normative decision theory is a pragmatic one. A reasonable point of departure when formulating descriptive hypotheses is that people behave rationally, at least most of the time. It would be difficult to reconcile the thought that most people most of the time make irrational decisions with the observation that they are in fact alive and seem to lead fairly good lives – in general, most of us seem to do pretty well. Moreover, if we were to discover that people actually behave irrationally, either occasionally or frequently, we would not be able to advise them how to change their behaviour unless we had some knowledge about normative decision theory. It seems that normative decision theory is better dealt with *before* we develop descriptive hypotheses.

That said, normative and descriptive decision theory share some common ground. A joint point of departure is that decisions are somehow triggered by the decision maker's beliefs and desires. This idea stems from the work of Scottish eighteenth-century philosopher David Hume. According to Hume, the best explanation of why Columbus set off westwards was that he *believed* it would be possible to reach the Far East by sailing in that direction, and that he *desired* to go there more than he desired to stay at home. Likewise, a possible explanation of why people bet in casinos is that they *believe* that the chance of winning large amounts is higher than it actually is and that they have a strong *desire* for money. In the twentieth century much work in descriptive decision theory was devoted to formulating mathematically precise hypotheses about how exactly beliefs and desires trigger choices. Unsurprisingly, a number of philosophers, economists and statisticians also proposed theories for how beliefs and desires *ought* to be aggregated into rational decisions.

1.2 Rational and right decisions

A decision can be rational without being right and right without being rational. This has been illustrated through many examples in history. For instance, in the battle of Narva (on the border between Russia and what we now call Estonia) on 20 November 1700, King Carl of Sweden and his 8,000 troops attacked the Russian army, led by Tsar Peter the Great. The tsar had about ten times as many troops at his disposal. Most historians agree that the Swedish attack was irrational, since it was almost certain to fail. Moreover, the Swedes had no strategic reason for attacking; they could not expect to gain very much from victory. However, because of an unexpected blizzard that blinded the Russian army, the Swedes won. The battle was over in less than two hours. The Swedes lost 667 men and the Russians approximately 15,000.

Looking back, the Swedes' decision to attack the Russian army was no doubt right, since the *actual outcome* turned out to be success. However, since the Swedes had no *good reason* for expecting that they were going to win the decision was nevertheless irrational. Decision theorists are primarily concerned with rational decisions, rather than right ones. In many cases it seems impossible to foresee, even in principle, which act is right until the decision has already been made (and even then it might be impossible to know what *would* have happened had one decided differently). It seems much more reasonable to claim that it is always possible to foresee whether a decision is rational. This is because theories of rationality operate on information available at the point in time the decision is made, rather than on information available at some later point in time.

More generally speaking, we say that a decision is *right* if and only if its actual outcome is at least as good as that of every other possible outcome. Furthermore, we say that a decision is *rational* if and only if the decision maker chooses to do what she has most reason to do at the point in time at which the decision is made. The kind of rationality we have in mind here is what philosophers call *instrumental* rationality. Instrumental rationality presupposes that the decision maker has some *aim*, such as becoming rich and famous, or helping as many starving refugees as possible. The aim is external to decision theory, and it is widely thought that an aim cannot in itself be irrational, although it is of course reasonable to think that *sets* of aims can sometimes be irrational, e.g. if they are mutually inconsistent. Now, on this view, to be instrumentally rational is to do whatever one has most reason to expect will fulfil one's aim. For instance, if your aim is not to get wet and it is raining heavily, you are rational in an instrumental sense if you bring an umbrella or raincoat when going for a walk.

The instrumental, means-to-end notion of rationality has been criticised, however. Philosopher John Rawls argues that an aim such as counting the number of blades of grass on a courthouse lawn is irrational, at least as long as doing so does not help to prevent terrible events elsewhere. Counting blades of grass on a courthouse lawn is not important enough to qualify as a rational aim. In response to this point it could perhaps be objected that everyone should be free to decide for herself what is important in life. If someone strongly desires to count blades of grass on courthouse lawns, just for the fun of it, that might very well qualify as a rational aim.

1.3 Risk, ignorance and uncertainty

In decision theory, everyday terms such as *risk*, *ignorance* and *uncertainty* are used as technical terms with precise meanings. In decisions under risk the decision maker knows the probability of the possible outcomes, whereas in decisions under ignorance the probabilities are either unknown or non-existent. Uncertainty is either used as a synonym for ignorance, or as a broader term referring to both risk and ignorance.

Although decisions under ignorance are based on less information than decisions under risk, it does not follow that decisions under ignorance must therefore be more difficult to make. In the 1960s, Dr Christiaan Barnard in Cape Town experimented on animals to develop a method for transplanting hearts. In 1967 he offered 55-year-old Louis Washkansky the chance to become the first human to undergo a heart transplant. Mr Washkansky was dying of severe heart disease and was in desperate need of a new heart. Dr Barnard explained to Mr Washkansky that no one had ever before attempted to transplant a heart from one human to another. It would therefore be meaningless to estimate the chance of success. All Dr Barnard knew was that his surgical method seemed to work fairly well on animals. Naturally, because Mr Washkansky knew he would not survive long without a new heart, he accepted Dr Barnard's offer. The donor was a 25-year-old woman who had died in a car accident the same day. Mr Washkansky's decision problem is illustrated in Table 1.2.

The operation was successful and Dr Barnard's surgical method worked quite well. Unfortunately, Mr Washkansky died 18 days later from pneumonia, so he did not gain as much as he might have hoped.

The decision made by Mr Washkansky was a decision under ignorance. This is because it was virtually impossible for him (and Dr Barnard) to assign meaningful probabilities to the possible outcomes. No one knew anything about the probability that the surgical method would work. However, it was nevertheless easy for Mr Washkansky to decide what to do. Because no matter whether the new surgical method was to work on humans or not, the outcome for Mr Washkansky was certain to be at least as good as if he decided to reject the operation. He had nothing to lose. Decision theorists say that in a case like this the first alternative (to have the operation)

Table 1.2

	Method works	Method fails
Operation	Live on for some time	Death
No operation	Death	Death

dominates the second alternative. The concept of dominance is of fundamental importance in decision making under ignorance, and it will be discussed in more detail in Chapter 3.

Since Mr Washkansky underwent Dr Barnard's pioneering operation, thousands of patients all over the world have had their lives prolonged by heart transplants. The outcomes of nearly all of these operations have been carefully monitored. Interestingly enough, the decision to undergo a heart transplant is no longer a decision under uncertainty. Increased medical knowledge has turned this kind of decision into a decision under risk. Recent statistics show that 71.2% of all patients who undergo a heart transplant survive on average 14.8 years, 13.9% survive for 3.9 years, and 7.8% for 2.1 years. However, 7.1% die shortly after the operation. To simplify the example, we shall make the somewhat unrealistic assumption that the patient's life expectancy after a heart transplant is determined entirely by his genes. We shall furthermore suppose that there are four types of genes.

Group I: People with this gene die on average 18 days after the operation (0.05 years).
Group II: People with this gene die on average 2.1 years after the operation.
Group III: People with this gene die on average 3.9 years after the operation.
Group IV: People with this gene die on average 14.8 years after the operation.

Since heart diseases can nowadays be diagnosed at a very early stage, and since there are several quite sophisticated drugs available, patients who decline transplantation can expect to survive for about 1.5 years. The decision problem faced by the patient is summarised in Table 1.3.

The most widely applied decision rule for making decisions under risk is the principle of maximising expected value. As will be explained in some detail in Chapter 4, this principle holds that the total value of an act equals the sum of the values of its possible outcomes weighted by the

Table 1.3

	Group I: 7.1%	Group II: 7.8%	Group III: 13.9%	Group IV: 71.2%
Operation	0.05 years	2.1 years	3.9 years	14.8 years
No operation	1.5 years	1.5 years	1.5 years	1.5 years

probability for each outcome. Hence, the expected values of the two alternatives are as follows.

Operation: $(0.05 \cdot 0.071) + (2.1 \cdot 0.078) + (3.9 \cdot 0.139)$
              $+(14.8 \cdot 0.712) \approx 11$
No operation: $(1.5 \cdot 0.071) + (1.5 \cdot 0.078) + (1.5 \cdot 0.139)$
              $+(1.5 \cdot 0.712) = 1.5$

Clearly, if the principle of maximising expected value is deemed to be acceptable, it follows that having an operation is more rational than not having one, since 11 is more than 1.5. Note that this is the case despite the fact that 7.1% of all patients die within just 18 days of the operation.

1.4 Social choice theory and game theory

The decisions exemplified so far are all decisions made by a *single* decision maker, *not* taking into account what other decision makers are doing. Not all decisions are like this. Some decisions are made collectively by a group, and in many cases decision makers need to take into account what others are doing. This has given rise to two important subfields of decision theory, viz. social choice theory and game theory.

Social choice theory seeks to establish principles for how decisions involving more than one decision maker ought to be made. For instance, in many countries (but unfortunately not all) political leaders are chosen by democratic election. Voting is one of several methods for making social choices. However, as will be explained in Chapter 13, the voting procedures currently used in many democratic countries are quite unsatisfactory from a theoretical perspective, since they fail to meet some very reasonable requirements that such procedures ought to fulfil. This indicates that the voting procedures we currently use may not be the best ones. By learning more about social choice theory we can eventually improve the way important decisions affecting all of us are made. Naturally, the group making a social choice need not always be the people of a nation; it could also be the members of, say, a golf club or a family. The basic theoretical problem is the same: How do we aggregate the divergent beliefs and desires of a heterogeneous set of individuals into a collective decision? In order to avoid misunderstanding, it is worth keeping in mind that collective entities, such as governments and corporations, sometimes act as single decision makers.

That is, not every act performed by a group is a social choice. For example, once the government has been elected its decisions are best conceived of as decisions taken by a single decision maker.

Game theory is another, equally important sub-field of decision theory. You are probably familiar with games such as chess and Monopoly, wherein the outcome of your decision depends on what others are doing. Many other decisions we make have the same basic structure, and if your opponents are clever enough they can foresee what you are likely to do, and then adjust their strategies accordingly. If you are rational, you will of course also adjust your strategy based on what you believe about your opponent. Here is an example, originally discussed by Jean-Jacques Rousseau: Two hunters can either cooperate to hunt a stag (which is a rather large animal that cannot be caught by a single hunter) or individually hunt for hares. A hare is rather small and can easily be caught by a single hunter. If the hunters cooperate and hunt stag each of them will get 25 kg of meat; this is the best outcome for both hunters. The worst outcome for each hunter is to hunt stag when the other is hunting hare, because then he will get nothing. If the hunter decides to hunt hare he can expect to get a hare of 5 kg. In Table 1.4 the numbers in each box refer to the amount of meat caught by the first and second hunter, respectively.

This game has become known as stag hunt. In order to analyse it, imagine that you are Hunter 1. Whether it would be better to hunt stag or hare depends on what you believe the other hunter will do. Note, however, that this also holds true for the other hunter. Whether it would be better for him to hunt stag or hare depends on what he believes you are going to do. If both of you were fully confident that the other would cooperate, then both of you would benefit from hunting stag. However, if only one hunter chooses to hunt stag and the other does not cooperate, the hunter will end

Table 1.4

		Hunter 2	
		stag	hare
Hunter 1	stag	25 kg, 25 kg	0 kg, 5 kg
	hare	5 kg, 0 kg	5 kg, 5 kg

up with nothing. If you were to hunt hare you would not have to worry about this risk. The payoff of hunting hare does not depend on what the other hunter chooses to do. The same point applies to the other hunter. If he suspects that you may not be willing to cooperate it is safer to hunt hare. Rational hunters therefore have to make a trade-off between two conflicting aims, viz. mutual benefit and risk minimisation. Each hunter is pulled towards stag hunting by considerations of mutual benefit, and towards hare hunting by considerations of risk minimisation. What should we expect two rational players to do when playing this game?

Many phenomena in society have a similar structure to the stag hunting scenario. In most cases we are all better off if we cooperate and help each other, but this cooperation can only occur if we trust our fellow citizens. Unfortunately, we sometimes have little or no reason to trust our fellow citizens. In such cases it is very likely that we will end up with outcomes that are bad for everyone. That said, there are of course also cases in which we tend to trust each other, even though the game has exactly the same structure as stag hunt. For instance, David Hume (1739: III) observed that, "Two men who pull at the oars of a boat, do it by an agreement or convention, tho' they have never given promises to each other." Arguably, the best outcome for both rowers is to cooperate, whereas the worst outcome is to row alone while the other is relaxing. Hence, from a game-theoretical point of view, stag hunting is similar to rowing. Why is it, then, that most people tend to cooperate when rowing but not when hunting stag? In Chapter 12 it will be explained that the answer has to do with the number of times the game is repeated.

Before closing this section, a note about terminology is called for. I – and many others – use the term *decision theory* both as a general term referring to all kinds of theoretical inquiries into the nature of decision making, including social choice theory and game theory, as well as a more narrow term referring only to individual decisions made by a single individual not considering the behaviour of others. Whether the term is used in the general or narrow sense is determined by context.

1.5 A very brief history of decision theory

The history of decision theory can be divided into three distinct phases: the Old period, the Pioneering period and the Axiomatic period. As is the case for nearly all academic disciplines, the Old period begins in ancient Greece. However, the Greeks did not develop a *theory* of rational decision making. They merely identified decision making as one of many areas worthy of further investigation. This is partly due to the fact that the Greeks were not familiar with the concept of probability. The Greek language contains expressions for everyday notions of likelihood and chance, but none of the great mathematicians working at that time formalised these ideas into a mathematical calculus. Furthermore, there is no evidence that Greek thinkers formulated any precise decision rules, at least not of the form discussed in contemporary decision theory. In Greek literature one can find practical recommendations about what to do and not to do, but these do not have the same general and abstract form as modern decision rules.

So how did the Greeks contribute to decision theory? To start with, we know that the Greeks were familiar with the distinction between right and rational acts, introduced in the beginning of this chapter. Herodotus, taking 'right' to mean rational and 'contrary to good counsel' to mean irrational, notes that,

> for even if the event turns out contrary to one's hope, still one's decision was right, even though fortune has made it of no effect: whereas if a man acts contrary to good counsel, although by luck he gets what he had no right to expect, his decision was not any the less foolish. (Herodotus VII: 10)

Furthermore, in Book III of his *Topics* Aristotle proposes what may best be conceived of as an embryo for the logic of rational preference. He says,

> if A be without qualification better than B, then also the best of the members of A is better than the best of the members of B; e.g. if Man be better than Horse, then also the best man is better than the best horse. (*Topics* III: ii)

This quote indicates that Aristotle was familiar with some of the problems of rational preference. However, the value of Aristotle's analysis is questioned by modern scholars. It has been pointed out that throughout his analysis Aristotle seems to mix logical principles, such as the one above, with contingent material principles. For example, Aristotle claims that, "to be a philosopher is better than to make money". This might very well be true, but it is merely a contingent truth, not a logical one. The decision to become a philosopher rather than a businessman is of little interest from a formal point of view; it appears to be nothing more than a piece of advice from a man who enjoyed being a philosopher.

The fifteen hundred years that followed the decline of the Greek culture is of little or no importance in the history of decision theory. There is no evidence that any significant advances in decision theory were made by the Romans or the early members of the Christian church. The second major development phase was the Pioneering period, which began in 1654 as Blaise Pascal and Pierre de Fermat began to exchange letters about probability theory. Their correspondence was triggered by a question asked by a French nobleman with an interest in gambling: What is the likelihood of getting at least one pair of sixes if a pair of fair dice is thrown 24 times? Rather than giving an approximate answer based on empirical trials, Fermat and Pascal started to work out an exact, mathematical solution to the problem. This led to the creation of modern probability theory. Christian Huygens, famous mainly for his work in astronomy and physics, learnt about the correspondence between Fermat and Pascal, and published the first book on the subject in 1657. (The answer to the nobleman's question is of course $1 - (35/36)^{24} \approx 0.49$.)

Another major breakthrough during the Pioneering period was the 1662 publication of a book known as *Port-Royal Logic*. Its real title was *La Logique, ou l'art de penser*. The book was published anonymously by two members of the Jansenist convent, an organisation belonging to the Catholic Church (which was later condemned by the church as heretical), working in Port-Royal just outside Paris. The main authors were Antoine Arnauld and Pierre Nicole, who were both associated with the convent, but it is widely believed that Pascal also contributed some portions of the text. *Port-Royal Logic* contains the first clear formulation of the principle of maximising expected value:

> In order to judge of what we ought to do in order to obtain a good and to avoid an evil, it is necessary to consider not only the good and evil in themselves, but also the probability of their happening and not happening, and to regard geometrically the proportion which all these things have, taken together. (Arnauld and Nicole 1662: 367)

Even though reasonable in many situations, it was soon observed that the principle of maximising expected value leads to counterintuitive recommendations if 'good' and 'evil' are interpreted as, for example, amounts of money. Given the choice between maintaining the status quo and a gamble in which one would either win or lose one million gold ducats with equal probability, *Port-Royal Logic* recommends us to be indifferent between the two options. Intuitively we know that cannot be right. In 1738 Daniel Bernoulli introduced the notion of *moral value* as an improvement on the imprecise terms 'good' and 'evil'. For Bernoulli, the moral value of an outcome is a technical term referring to how good or bad that outcome is from the decision maker's point of view. For example, a multi-millionaire might be indifferent between the status quo and a gamble in which he will either win or lose a million, simply because he is already very well off. For the rest of us this is unlikely to be the case. Since I am not a multi-millionaire, the negative moral value of losing a million is for me much greater than the positive moral value of winning a million. In the modern literature the term 'moral value' has been replaced by the notion of utility, but the meaning of the two terms is the same.

Modern decision theory is dominated by attempts to axiomatise the principles of rational decision making, which explains why we call it the Axiomatic period. This period had two separate and unrelated starting points. The first was Frank Ramsey's paper 'Truth and probability', written in 1926 but published posthumously in 1931. Ramsey was a philosopher working at Cambridge together with Russell, Moore and Wittgenstein. In his paper on probability, he proposed a set of eight axioms for how rational decision makers ought to choose among uncertain prospects. He pointed out that every decision maker behaving in accordance with these axioms will act in a way that is *compatible* with the principle of maximising expected value, by implicitly assigning numerical probabilities and values to outcomes. However, it does not follow that the decision maker's choices were *actually triggered* by these implicit probabilities and utilities. This way of thinking about rational decision making is very influential in the modern literature, and will be explained in more detail in Chapters 5, 6 and 7.

The second starting point was von Neumann and Morgenstern's book *Theory of Games and Economic Behavior*, and in particular the publication of the second edition in 1947. Von Neumann and Morgenstern were not aware of Ramsey's paper when the first edition of the book was published in 1944. The first chapter of the first edition was devoted to individual decision making under risk, and some people complained that von Neumann and

Morgenstern did not give any technically precise definition of the notion of utility (that is, their term for Bernoulli's concept of moral value) to which they frequently referred. Therefore, in the second edition they responded by presenting a set of axioms for how rational decision makers ought to choose among lotteries. For von Neumann and Morgenstern a lottery is a probabilistic mixture of outcomes; for example, 'a fifty-fifty chance of winning either $100 or a ticket to the opera' is a lottery. They showed that every decision maker behaving in accordance with their axioms implicitly behaves in accordance with the principle of maximising expected utility, and implicitly assigns numerical utilities to outcomes. The main difference compared to Ramsey's axiomatisation is that von Neumann and Morgenstern presented no novel theory of probability. Nevertheless, as their work on game theory was genuinely new, it must be regarded as a major breakthrough.

The 1950s turned out to be a golden age for decision theory. Several influential papers and books were published during that period, including Leonard Savage's 1954 book *The Foundations of Statistics*, in which he presented yet another influential axiomatic analysis of the principle of maximising expected utility. A surprisingly large number of the books and papers published during those days are still widely read and quoted, and anyone wishing to understand the latest developments in decision theory is usually well advised to first read about how similar problems were analysed in the 1950s.

Exercises

1.1 Explain the difference between:
   - (a) Decisions under risk and ignorance
   - (b) Social choices and decisions made by individuals
   - (d) Games and all other types of decisions discussed here
1.2 Consider the following situations:
   - (a) Charles Lindbergh was the first man to fly single-handed across the Atlantic in 1927. Did he make a decision under risk or ignorance as he departed New York and set off eastwards?
   - (b) You are thinking about flying to Paris next week. Are you making a decision under risk or ignorance?
1.3 Consider the four lotteries below. Exactly one winning ticket will be drawn in each lottery.

	Ticket no. 1	Ticket no. 2–20	Ticket no. 21–100
Lottery A	$2 million	$2 million	$2 million
Lottery B	$0	$15 million	$2 million
Lottery C	$2 million	$2 million	$0
Lottery D	$0	$15 million	$0

(a) Make intuitive comparisons between lottery A and B, and between C and D, without performing any calculations. Do you prefer A or B? C or D?

(b) Now calculate the expected (monetary) value of each lottery. Which lottery has the highest expected monetary value, A or B? C or D?

(c) Did these calculations make you change your mind? If so, why? (This is a version of Allais' paradox, which we will discuss in more detail in Chapter 4.)

1.4 Consider the following game, in which both you and your opponent have two alternatives to choose between. The first number in each box represents your payoff, whereas the second number represents your opponent's payoff. (Naturally, better payoffs are represented by higher numbers.)

		Your Opponent	
		Alt 1	Alt 2
You	Alt 1	1, 1	0, 3
	Alt 2	3, 0	2, 2

(a) What do you expect your opponent to do?

(b) What will you do?

(c) Try to explain why this game is of less theoretical interest than the stag hunt game.

1.5 Briefly summarise the major events in the history of decision theory.

Solutions

1.1 Explicit answers can be found in the text.

1.2 (a) It depends on what he knew when he took off from New York, but it is reasonable to assume that he could not assign any probabilities to the various outcomes of his trip; if so, he made a decision under ignorance.

(b) You are making a decision under risk. Even if you do not know the probability of a crash right here and now, there is certainly a lot of relevant information available in libraries and on the Internet.

1.3 (a) Empirical studies by psychologists show that most people prefer A over B, and D over C.

(b) The expected monetary values are: A = \$2M, B = \$4.45M, C = \$0.4M, D = \$2.85M.

(c) Yes, if you accept the principle of maximising expected monetary value. But is this really a reasonable decision rule? See Chapter 4!

1.4 (a) Your opponent will choose alternative 2 (no matter what you do).

(b) You will, if you are rational, also choose alternative 2 (no matter what your opponent does).

(c) It is less interesting than Stag Hunt simply because there is no conflict between cooperation and other aims; alternative 2 dominates alternative 1, in the sense that one is better off choosing alternative 2 no matter what the other player does.

1.5 An explicit answer can be found in the text.

2 The decision matrix

Before you make a decision you have somehow to determine what to decide about. Or, to put it differently, you have to specify what the relevant *acts*, *states* and *outcomes* are. Suppose, for instance, that you are thinking about taking out fire insurance on your home. Perhaps it costs $100 to take out insurance on a house worth $100,000, and you ask: Is it worth it? Before you decide, you have to get the formalisation of the decision problem right. In this case, it seems that you face a decision problem with two acts, two states, and four outcomes. It is helpful to visualise this information in a decision matrix; see Table 2.1.

To model one's decision problem in a formal representation is essential in decision theory, since decision rules are only defined relative to such formalisations. For example, it makes no sense to say that the principle of maximising expected value recommends one act rather than another unless there is a formal listing of the available acts, the possible states of the world and the corresponding outcomes. However, instead of visualising information in a decision matrix it is sometimes more convenient to use a decision tree. The decision tree in Figure 2.1 is equivalent to the matrix in Table 2.1.

The square represents a *choice node*, and the circles represent *chance nodes*. At the choice node the decision maker decides whether to go *up* or *down* in the tree. If there are more than two acts to choose from, one simply adds more lines. At the chance nodes nature decides which line to follow, and the rightmost boxes represent the possible outcomes. Decision trees are often used for representing sequential decisions, i.e. decisions that are divided into several separate steps. (Example: In a restaurant, you can either order all three courses before you start to eat, or divide the decision-making process into three separate decisions taken at three points in time. If you opt for the latter approach, you face a sequential decision problem.) To

Table 2.1

	Fire	No fire
Take out insurance	No house and $100,000	House and $0
No insurance	No house and $100	House and $100

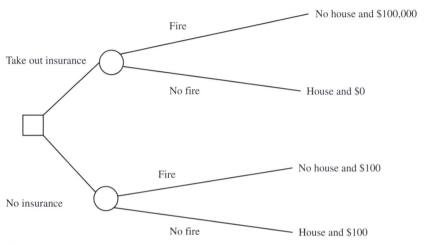

Figure 2.1

represent a sequential decision problem in a tree, one simply adds new choice and chance nodes to the right of the existing leafs.

Many decision theorists distinguish only between decision problems and a corresponding decision matrix or tree. However, it is worth emphasising that we are actually dealing with three levels of abstraction:

1. The decision problem
2. A formalisation of the decision problem
3. A visualisation of the formalisation

A decision problem is constituted by the entities of the world that prompt the decision maker to make a choice, or are otherwise relevant to that choice. By definition, a formalisation of a decision problem is made up of information about the decision to be made, irrespective of how that information is visualised. Formalisations thus comprise information about acts, states and outcomes, and sometimes also information about probabilities. Of course, one and the same decision problem can be formalised in

Table 2.2

[

[a_1 = take out insurance,

a_2 = do not];

[s_1 = fire,

s_2 = no fire];

[(a_1, s_1) = No house and $100,000,

(a_1, s_2) = House and $0,

(a_2, s_1) = No house and $100,

(a_2, s_2) = House and $100]

]

---

different ways, not all of which are likely to be equally good. For example, some decision problems *can* be formalised either as decisions under risk or as decisions under ignorance, but if probabilities are known it is surely preferable to choose the former type of formalisation (since one would otherwise overlook relevant information).

Naturally, any given set of information can be visualised in different ways. We have already demonstrated this by drawing a matrix and a tree visualising the same formalisation. Table 2.2 is another example of how the same information could be presented, which is more suitable to computers.

In Table 2.2 information is stored in a *vector*, i.e. in an ordered list of mathematical objects. The vector is comprised of three new vectors, the first of which represents acts. The second vector represents states, and the third represents outcomes defined by those acts and states.

From a theoretical perspective, the problem of how to formalise decision problems is arguably more interesting than questions about how to visualise a given formalisation. Once it has been decided what pieces of information ought to be taken into account, it hardly matters for the decision theorist whether this information is visualised in a matrix, a tree or a vector.

2.1 States

The basic building blocks of a decision problem are states, outcomes and acts. Let us discuss each concept in turn, starting with states. What is a state? Intuitively, a state is a part of the world that is not an outcome or an act (that can be performed by the agent in the present decision situation;

Table 2.3

	I choose the right bet	I do not
Bet on Democrat	$100	$0
Bet on Republican	$200	$0

acts performed by others can presumably be thought of as states). It is difficult to come up with a more precise definition without raising deep metaphysical questions that fall well beyond the scope of this book.

That said, not all states are relevant to decision making. For many decisions it is completely irrelevant whether the number of atoms in the universe is odd or even, for instance. Only states that may affect the decision maker's preference among acts need to be taken into account, such as: *The republican candidate wins the election*, or *The interest rate exceeds five per cent next year*, or *My partner loves me*, or *Goldbach's conjecture is true*. For each of these states, we can easily imagine an act whose outcome depends on the state in question. The example of Goldbach's conjecture (a famous mathematical hypothesis) indicates that even necessary truths may be relevant in decision making, e.g. if the decision maker has placed a bet on whether this hypothesis is true or not.

Some states, or at least some ways of *partitioning* states, are clearly illegitimate. In order to understand why, imagine that you are offered a choice between two bets, which pay $100 and $200 respectively, depending on whether the Democrat or the Republican candidate will win the next presidential election (Table 2.3). Now, it would make little sense to consider the states *I choose the right bet* and *I do not*.

This formalisation gives the false impression that you will definitely be better off if you choose to bet on the Republican candidate. The reason why this is false, and why the formalisation is illegitimate, is that the state *I choose the right bet* is causally dependent of the act you choose. Whether the state will occur depends on which act is chosen.

The problem of causal dependence can be addressed in two ways. The first is to allow the decision theorist to include only states that are causally independent of the acts in the formalisation; this is the option that we shall pursue here (with exception for Chapter 9). The second option is to avoid the notion of states altogether. That approach works particularly well if one

Table 2.4

Bet on Democrat	$100 (0.6)	$0 (0.4)
Bet on Republican	$200 (0.3)	$0 (0.7)

Table 2.5(a)

	State 1	State 2
Act A	$50	$80
Act B	$30	$80

Table 2.5(b)

Act A	$50	$80
Act B	$80	$30

happens to know the probability of the outcomes. Suppose, for instance, that the probability is 0.6 that you win $100 if you bet on the Democrat, and 0.3 that you win $200 if you bet on the Republican.

Arguably, the formalisation in Table 2.4 is impeccable. That said, omitting the states makes little sense in other decision problems, such as decisions under ignorance. If you do not know anything about the probability that your house will burn down and you are offered fire insurance for free, it certainly makes sense to accept the offer. No matter what happens, you will be at least as well off if you accept free insurance than if you do not, and you will be better off if there is a fire. (This line of reasoning has a fancy name: *the dominance principle*. See Chapter 3.) However, this conclusion only follows if we attach outcomes to states. The pair of examples in Table 2.5(a) and 2.5(b) illustrates the difference between including and omitting a set of states.

In Table 2.5(a), act A is clearly better than act B. However, in Table 2.5(b) all states have been omitted and the outcomes have therefore been listed in *arbitrary* order. Here, we fail to see that one option is actually better than the other. This is why it is a bad idea not to include states in a formalisation.

Let us now consider a slightly different kind of problem, having to do with how preferences depend on which state is in fact the true state. Suppose that you have been severely injured in a traffic accident and that

Table 2.6

	You survive	You die
Take first bet	$100	$0
Take second bet	$0	$100

as a result you will have to undergo a risky and life-threatening operation. For some reason, never mind why, you are offered a choice between a bet in which you win $100 if the operation is successful and you survive, and nothing if the operation fails and you die, and a bet in which you win nothing if the operation is successful and you survive, and $100 if the operation fails and you die (Table 2.6).

Let us suppose that both states are equally probable. Then, it is natural to argue that the decision maker should regard both bets as equally attractive. However, most people would prefer the bet in which one wins $100 if one survives the operation, no matter how improbable that state is. If you die, money does not matter to you any more. This indicates that the formalisation is underspecified. To make the formalisation acceptable we would have to add to the outcome the fact that the decision maker will die if the operation fails. Then, it would clearly transpire why the first bet is better than the second. To put it in a more sophisticated way, to which we shall return in Chapter 7, states should be chosen such that the value of the outcomes under all states is independent of whether the state occurs or not.

2.2 Outcomes

Rational decision makers are not primarily concerned with states or acts. What ultimately matters is the *outcome* of the choice process. Acts are mere instruments for reaching good outcomes, and states are devices needed for applying these instruments. However, in order to figure out which instrument to use (i.e. which act to choose given a set of states), outcomes must be ranked in one way or another, from the worst to the best. Exactly how this should be done is an important topic of debate in decision theory, a topic which we shall examine in more detail in Chapter 5. In the present section we shall merely explain the difference between the various kinds of scales that are used for comparing outcomes.

Let us return to the issue of whether or not one should insure a house worth $100,000 at a rate of $100 per annum. Imagine that Jane has made a sincere effort to analyse her attitudes towards safety and money, and that she felt that the four possible outcomes should be ranked as follows, from the best to the worst.

1. House and $100   *is better than*
2. House and $0   *is better than*
3. No house and $100,000   *is better than*
4. No house and $100.

The first outcome, 'House and $100', can be thought of as a possible world that is exactly similar to the three others, except for the condition of the house and amount of money in that world. Outcomes are in that sense *comprehensive* – they include much more than we actually need to mention in a decision matrix. Naturally, the ranking of outcomes is to a large extent subjective. Other decision makers may disagree with Jane and feel that the outcomes ought to be ranked differently. For each decision maker, the ranking is acceptable only if it reflects his or her attitudes towards the outcomes. It seems fairly uncontroversial to suppose that people sometimes have different attitudes, but the ranking depicted above does, we assume, accurately reflect Jane's attitudes.

In order to measure the value of an outcome, as it is perceived by the decision maker, it is convenient to assign numbers to outcomes. In decision theory, numbers referring to comparative evaluations of value are commonly called *utilities*. However, the notion of utility has many different technical meanings, which should be kept separate. Therefore, to avoid unnecessary confusion we shall temporarily stick to the rather vague term *value*, until the concept of utility has been properly introduced in Chapter 5.

Value can be measured on two fundamentally different kinds of scales, viz. *ordinal* scales and *cardinal* scales. Consider the set of numbers in Table 2.7, assigned by Jane to the outcomes of her decision problem.

Table 2.7

	Fire	No fire
Take out insurance	1	4
Do not	−100	10

Table 2.8

	Original scale	Scale A	Scale B	Scale C
Best outcome	10	4	100	777
Second best	4	3	98	−378
Third best	1	2	97	−504
Worst outcome	−100	1	92	−777

If Jane assigns a higher number to one outcome than another, she judges the first outcome to be better than the second. However, if the scale she uses is an ordinal scale, nothing more than that follows. In particular, nothing can be concluded about *how much* better one outcome is in relation to another. The numbers merely reflect the qualitative ranking of outcomes. No quantitative information about the 'distance' in value is reflected by the scale. Look at Scales A–C in Table 2.8; they could be used for representing exactly the same ordinal ranking.

The transformations of the original scale into scales A, B or C preserves the order between the outcomes. This proves that all four scales are equivalent. Hence, it does not matter which set of numbers one uses. Mathematicians express this point by saying that ordinal scales are *invariant* up to *positive monotone transformations*. That a transformation of a scale is invariant under some sort of change means that the ranking of the objects is preserved after this type of change. In the case of an ordinal scale, the change is describable by some function f such that

$$f(x) \geq f(y) \text{ if and only if } x \geq y. \tag{1}$$

In this expression, x and y are two arbitrary values of some initial scale, e.g. the values corresponding to the best and second best outcomes on the original scale above, and f is some mathematical function. It can be easily verified that the transformation of the initial scale into scale D in Table 2.9 *does not* satisfy condition (1), because if $x = 10$ and $y = 4$, then $f(x) = 8$, and $f(y) = 9$. Of course, it is false that $8 \geq 9$ if and only if $10 \geq 4$. It can also be shown in analogous ways that scales E and F are not permissible ordinal transformations of the original scale.

As pointed out above, ordinal scales are usually contrasted with cardinal scales. Cardinal scales embody more information than ordinal scales. There are two different kinds of cardinal scales, viz. *interval* scales and *ratio* scales.

Table 2.9

	Scale D	Scale E	Scale F
Best outcome	8	−60	100
Second best	9	−50	90
Third best	6	−40	80
Worst outcome	7	0	80

To start with, we focus on interval scales. Unlike ordinal scales, interval scales accurately reflect the difference between the objects being measured. Let us suppose, for illustrative purposes, that scale F in the example above is an interval scale. It would then be correct to conclude that the difference in value between the best and the second best outcome is exactly the same as the distance in value between the second best and the third best outcome. Furthermore, the difference between the best and the worst outcome is twice that between the best and the second best outcome. However, scale E cannot be a permissible transformation of F, because in F the distance between the third best and the worst outcome is zero, whereas the corresponding difference in E is strictly greater than zero.

To illustrate what kind of information is represented in an interval scale, consider the two most frequently used scales for measuring temperature, i.e. the Centigrade (C) and Fahrenheit (F) scales, respectively. Both scales accurately reflect differences in temperature, and any temperature measured on one scale can easily be transformed into a number on the other scale. The formula for transforming Centigrade to Fahrenheit is:

$$F = 1.8 \cdot C + 32 \tag{2}$$

By solving this equation for C, we get:

$$C = (F - 32)/1.8 \tag{3}$$

Note that (2) and (3) are straight lines – had the graphs been curved, a difference of, say, one unit on the x-axis will not always produce the same difference on the y-axis. As an illustration of how the Fahrenheit scale can be transformed into the centigrade scale, consider Table 2.10. It shows the temperatures in a number of cities on a sunny day a few years ago.

When looking at Table 2.10, a common error is to conclude that it was *twice as warm* in Tokyo as in New York, since 64 units Fahrenheit is twice as

Table 2.10

City	Degrees Fahrenheit	Degrees Centigrade
Los Angeles	82	27.8
Tokyo	64	17.8
Paris	62	16.7
Cambridge (UK)	46	7.8
New York	32	0
Stockholm	−4	−20

much as 32 units Fahrenheit. In order to see why that conclusion is incorrect, note that 64 °F corresponds to 17.8 °C and 32 °F to 0 °C. Now, 17.8 °C is of course not twice as much as 0 °C. Had it been twice as warm in Tokyo as in New York, it would certainly have been twice as warm according to every scale. Interval scales accurately reflect differences, but not ratios. Expressed in mathematical terminology, interval scales are invariant up to *positive linear transformations*. This means that any interval scale can be transformed into another by multiplying each entry by a positive number and adding a constant, without losing or gaining any information about the objects being measured. For example, if the value of some outcome is 3 according to scale X, and $Y = 10 \cdot X + 5$, then the value of the same outcome would be 35 if measured on scale Y. Obviously, scale Y is obtained from X by a positive linear transformation.

Unlike interval scales, ratio scales accurately reflect ratios. Mass, length and time are all examples of entities that can be measured on ratio scales. For example, 20 lb = 9 kg, and this is twice as much as 10 lb = 4.5 kg. Furthermore, two weeks is twice as much as one, and 14 days is of course twice as much as 7 days. Formally put, a ratio scale U can be accurately transformed into an equivalent ratio scale V by multiplying U by a positive constant k. Consider, for instance, the series of numbers in Table 2.11, which denote the values of four different outcomes as measured on five different scales, G–K.

Scale G and I are equivalent ratio scales, because $G = 5 \cdot I$. Furthermore, the first four scales, G–J, are equivalent interval scales. For example, scale H can be obtained from G by the formula $G = 10 \cdot H + 10$, and $J = 0.1 \cdot G - 3$. However, there is no equation of the form $V = k \cdot U + m$ that transforms G into K. Hence, since K is not a positive linear transformation of G, it does not reflect the same differences in value, nor the same ratios.

Table 2.11

	Scale G	Scale H	Scale I	Scale J	Scale K
Best outcome	40	410	8	1	5
Second best	30	310	6	0	3
Third best	20	210	4	−1	2
Worst	10	110	2	−2	1

It is helpful to summarise the technical properties of the two kinds of cardinal scales discussed here in two mathematical conditions. To start with, a function f that takes an argument x and returns a real number as its value is an interval scale if and only if condition (1) on page 24 holds *and* for every other function f' that satisfies (1) there are some positive constants k and m such that:

$$f'(x) = k \cdot f(x) + m \tag{4}$$

Condition (4) states what transformations of an interval scale are permissible: As we have shown above, every transformation that can be mapped by an upward sloping straight line is permissible. Furthermore, a function f that takes an argument x and returns a real number as its value is a ratio scale if and only if condition (1) holds *and* for every other function f' that satisfies (1) there is some positive constant k such that:

$$f'(x) = k \cdot f(x) \tag{5}$$

This condition is even more simple that the previous one: A pair of ratio scales are equivalent if and only if each can be transformed into the other by multiplying all values by some positive constant. (Of course, the constant we use for transforming f into f' is not the same as that we use for transforming f' into f.)

2.3 Acts

Imagine that your best friend Leonard is about to cook a large omelette. He has already broken five eggs into the omelette, and plans to add a sixth. However, before breaking the last egg into the omelette, he suddenly starts to worry that it might be rotten. After examining the egg carefully, he decides to take a chance and break the last egg into the omelette.

Box 2.1   Three types of scales

In this chapter we have discussed three different types of scales. Their main characteristics can be summarised as follows.

1. Ordinal scale: Qualitative comparison of objects allowed; no information about differences or ratios. Example: The jury of a song contest award points to the participants. On this scale, 10 points is more than 5.

2. Cardinal scales

  (a) Interval scale   Quantitative comparison of objects; accurately reflects differences between objects. Example: The Centigrade and Fahrenheit scales for temperature measurement are the most well-established examples. The difference between 10 °C and 5 °C equals that between 5 °C and 0 °C, but the difference between 10 °C and 5 °C does not equal that between 10 °F and 5 °F.

  (b) Ratio scale   Quantitative comparison of objects; accurately reflects ratios between objects. Example: Height, mass, time, etc. 10 kg is twice as much as 5 kg, and 10 lb is also twice as much as 5 lb. But 10 kg is not twice as much as 5 lb.

The act of adding the sixth egg can be conceived of as a function that takes either the first state (*The sixth egg is rotten*) or the second (*The sixth egg is not rotten*) as its argument. If the first state happens to be the true state of the world, i.e. if it is inserted into the function, then it will return the outcome *No omelette*, and if the second state happens to be the true state, the value of the function will be *Six egg omelette*. (See Table 2.12.) This definition of acts can be trivially generalised to cover cases with more than two states and outcomes: an act is a function from a set of states to a set of outcomes.

Did you find this definition too abstract? If so, consider some other function that you are more familiar with, say $f(x) = 3x + 8$. For each argument x, the function returns a value $f(x)$. Acts are, according to the suggestion above and originally proposed by Leonard Savage, also functions. However, instead of taking numbers as their arguments they take states, and instead of returning other numbers they return outcomes. From a mathematical

Table 2.12

	The sixth egg is rotten	The sixth egg is not rotten
Add sixth egg	No omelette	Six egg omelette
Do not add sixth egg	Five egg omelette	Five egg omelette

point of view there is nothing odd about this; a function is commonly defined as any device that takes one object as its argument and returns exactly one other object. Savage's definition fulfils this criterion. (Note that it would be equally appropriate to consider states and acts as primitive concepts. Outcomes could be conceived of as ordered pairs of acts and states. For example, the outcome *No omelette* is the ordered pair comprising of the act *Add sixth egg* and the state *The sixth egg is rotten*. States can be defined in similar ways, in terms of acts and outcomes.)

Decision theory is primarily concerned with *particular* acts, rather than generic acts. A *generic* act, such as sailing, walking or swimming can be instantiated by different agents at different time intervals. Hence, Columbus' first voyage to America and James Cook's trip to the southern hemisphere are both instantiations of the same generic act, viz. sailing. Particular acts, on the other hand, are always carried out by specific agents at specific time intervals, and hence Columbus' and Cook's voyages were different particular acts. Savage's definition is a characterisation of particular acts.

It is usually assumed that the acts considered by a decision maker are *alternative* acts. This requirement guarantees that a rational decision maker has to choose only one act. But what does it mean to say that some acts constitute a set of alternatives? According to an influential proposal, the set A is an *alternative-set* if and only if every member of A is a particular act, A has at least two different members, and the members of A are agent-identical, time-identical, performable, incompatible in pairs and jointly exhaustive. At first glance, these conditions may appear as fairly sensible and uncontroversial. However, as pointed out by Bergström (1966), they do not guarantee that every act is a member of only one alternative-set. Some particular acts are members of several non-identical alternative-sets. Suppose, for instance, that I am thinking about going to the cinema (act a_1) or not going to the cinema (a_2), and that $\{a_1, a_2\}$ is an alternative-set. Then I realise that a_1 can be performed in different ways. I can, for instance, buy popcorn at the cinema (a_3) or buy chocolate (a_4). Now, also {a_1 & a_3, a_1 & a_4, a_2} is an alternative-set. Of course, a_1 & a_3 and a_1 & a_4 are different particular acts, so both of them cannot be identical to a_1. Moreover, a_1 & a_3 can also be performed in different ways. I can buy a small basket of popcorn (a_5) or a large basket (a_6), and therefore {a_1 & a_3 & a_5, a_1 & a_3 & a_6, a_1 & a_4, a_2} also constitutes an alternative-set, and so on and so forth. So what are the alternatives to a_2? Is it {a_1}, or {a_1 & a_3, a_1 & a_4, a_2}, or {a_1 & a_3 & a_5, a_1 & a_3 & a_6, a_1 & a_4, a_2}? Note that nothing excludes that the outcome of a_2 is better than the outcome of a_1 & a_3, while the outcome of a_1 & a_3 & a_6 might be better than that of a_2. This obviously causes problems for decision makers seeking to achieve as good outcomes as possible.

The problem of defining an alternative-set has been extensively discussed in the literature. Bergström proposed a somewhat complicated solution of the problem, which has been contested by others. We shall not explain it here. However, an interesting implication of Bergström's proposal is that the problem of finding an alternative-set is partly a normative problem. This is because we cannot formalise a decision problem until we know which normative principle to apply to the resolution of the problem. What your alternatives are depends partly on what your normative principle tells you to seek to achieve.

2.4 Rival formalisations

In the preceding sections, we have briefly noted that one cannot take for granted that there exists just one *unique* best formalisation of each decision problem. The decision maker may sometimes be confronted with *rival* formalisations of one and the same decision problem. Rival formalisations arise if two or more formalisations are equally reasonable and strictly better than all alternative formalisations.

Obviously, rival formalisations are troublesome if an act is judged to be rational in one optimal formalisation of a decision problem, but non-rational in another optimal formalisation of the same decision problem. In such cases one may legitimately ask whether the act in question should be performed or not. What should a rational decision maker do? The scope of this problem is illustrated by the fact that, theoretically, there might be cases in which *all* acts that are rational in one optimal formalisation are non-rational in another rival formalisation of the same decision problem, whereas *all* acts that are rational according to the latter formalisation are not rational according to the former.

To give convincing examples of rival formalisations is difficult, mainly because it can always be questioned whether the suggested formalisations are equally reasonable. In what follows we shall outline a hypothetical example that some people may find convincing, although others may disagree. Therefore, in Box 2.2 we also offer a more stringent and technical argument that our example actually is an instance of two equally reasonable but different formalisations.

Imagine that you are a paparazzi photographer and that rumour has it that actress Julia Roberts will show up in either New York (NY), Los Angeles (LA) or Paris (P). Nothing is known about the probability of these states of the world. You have to decide if you should stay in America or catch a plane to Paris. If you stay and actress Julia Roberts shows up in Paris you get $0; otherwise you get your photos, which you will be able to sell for $10,000. If you catch a plane to Paris and Julia Roberts shows up in Paris your net gain after having paid for the ticket is $5,000, and if she shows up in America you for some reason, never mind why, get $6,000. Your initial representation of the decision problem is visualised in Table 2.13.

Since nothing is known about the probabilities of the states in Table 2.13, you decide it makes sense to regard them as equally probable, i.e. you decide to assign probability 1/3 to each state. Consider the decision matrix in Table 2.14.

Table 2.13

	P	LA	NY
Stay	$0	$10k	$10k
Go to Paris	$5k	$6k	$6k

Table 2.14

	P (1/3)	LA (1/3)	NY (1/3)
Stay	$0	$10k	$10k
Go to Paris	$5k	$6k	$6k

Table 2.15

	P (1/3)	LA or NY (2/3)
Stay	$0	$10k
Go to Paris	$5k	$6k

Table 2.16

	P	LA or NY
Stay	$0	$10k
Go to Paris	$5k	$6k

Table 2.17

	P (1/2)	LA or NY (1/2)
Stay	$0	$10k
Go to Paris	$5k	$6k

The two rightmost columns are exactly parallel. Therefore, they can be merged into a single (disjunctive) column, by adding the probabilities of the two rightmost columns together (Table 2.15).

However, now suppose that you instead start with Table 2.13 and *first* merge the two repetitious states into a single state. You would then obtain the decision matrix in Table 2.16.

Now, since you know nothing about the probabilities of the two states, you decide to regard them as equally probable, i.e. you assign a probability of 1/2 to each state. This yields the formal representation in Table 2.17, which is clearly different from the one suggested above in Table 2.15.

Which formalisation is best, 2.15 or 2.17? It seems question begging to claim that one of them must be better than the other – so perhaps they are equally reasonable? If they are, we have an example of rival formalisations.

Note that the principle of maximising expected value recommends different acts in the two matrices. According to Table 2.15 you should stay, but 2.17 suggests you should go to Paris. Arguably, this example shows that rival formalisations must be taken seriously by decision theorists, although there is at present no agreement in the literature on how this phenomenon ought to be dealt with.

Box 2.2   Why rival representations are possible

The examples illustrated in Tables 2.15 and 2.17 do not *prove* that rival formalisations are possible. One may always question the claim that the two formalisations are equally reasonable. Therefore, in order to give a more comprehensive argument for thinking that the formalisations are equally reasonable, we shall introduce some technical concepts. To begin with, we need to distinguish between two classes of decision rules, viz. *transformative* and *effective* decision rules. A decision rule is effective if and only if it singles out some set of recommended acts, whereas it is transformative if and only if it modifies the formalisation of a given decision problem. Examples of effective decision rules include the principle of maximising expected utility and the dominance principle, mentioned in Chapter 1. Transformative decision rules do not directly recommend any particular act or set of acts. Instead, they transform a given formalisation of a decision problem into another by adding, deleting or modifying information in the initial formalisation. More precisely, transformative decision rules can alter the set of alternatives or the set of states of the world taken into consideration, modify the probabilities assigned to the states of the world, or modify the values assigned to the corresponding outcomes. For an example of a transformative decision rule, consider the rule saying that if there is no reason to believe that one state of the world is more probable than another then the decision maker should transform the initial formalisation of the decision problem into one in which every state is assigned equal probability. This transformative rule is called the principle of insufficient reason.

We assume that all significant aspects of a decision problem can be represented in a triplet $\pi = \langle A, S, O \rangle$, where A is a non-empty set of (relevant) alternative acts, S is a non-empty set of states of the world, and O is a set of outcomes. Let us call such a triplet a *formal decision problem*. A transformative decision rule is defined as a function t that transforms one formal decision problem π into another π', i.e. t is a transformative decision rule in a set of formal decision problems Π if and only if t is a function such that for all $\pi \in \Pi$, it holds that $t(\pi) \in \Pi$. If t and u form a pair of transformative decision rules, we can construct a new composite rule $(t \circ u)$ such that $(t \circ u)(\pi) = u(t(\pi))$. In this framework the question, "How should the decision maker formalise a decision problem?" can be restated as: "What sequence of transformative rules $(t \circ u \circ \ldots)$ should a rational decision maker apply to an initial formal decision problem π?".

Let $\succeq$ be a relation on Π such that $\pi \succeq \pi'$ if and only if the formal representation π is at least as reasonable as π'. (If π and π' are equally reasonable we write $\pi \sim \pi'$.) We shall not go into detail here about what makes one formal representation more reasonable than another, but it should be obvious that some representations are better than others. Now, we shall prove that if the technical condition stated below holds for $\succeq$, then two different sequences of a given set of transformative decision rules, $(t \circ u)$ and $(u \circ t)$, will always yield formalisations that are equally reasonable.

Order-independence (OI): $(u \circ t)(\pi) \succeq t(\pi) \succeq (t \circ t)(\pi)$

The left-hand inequality, $(u \circ t)(\pi) \succeq t(\pi)$, states that a transformative rule u should not, metaphorically expressed, throw a spanner in the works carried out by another rule t. Hence, the formalisation obtained by first applying u and then t has to be at least as good as the formalisation obtained by only applying t. The right-hand inequality, $t(\pi) \succeq (t \circ t)(\pi)$, says that nothing can be gained by immediately repeating a rule. This puts a substantial constraint on transformative rules; only 'maximally efficient' rules, that directly improve the formal representation as much as possible, are allowed by the OI-condition. Now consider the following theorem.

Theorem 2.1 Let the OI-condition hold for all $\pi \in \Pi$. Then, $(u \circ t)(\pi) \sim (t \circ u)(\pi)$ for all u and t.

Proof We prove Theorem 2.1 by making a series of substitutions:

(1) $(u \circ t \circ u)(\pi) \succeq (t \circ u)(\pi)$ Substitute $t \circ u$ for t in OI

(2) $(u \circ t \circ u \circ t)(\pi) \succeq (t \circ u)(\pi)$ From (1) and OI, substitute $t(\pi)$ for π

(3) $(u \circ t)(\pi) \succeq (t \circ u)(\pi)$ Right-hand side of OI

(4) $(t \circ u \circ t)(\pi) \succeq (u \circ t)(\pi)$ Substitute $u \circ t$ for t and u for t in OI

(5) $(t \circ u \circ t \circ u)(\pi) \succeq (u \circ t)(\pi)$ From (4) and OI, substitute t for u and u for t in OI, then substitute $u(\pi)$ for π

(6) $(t \circ u)(\pi) \succeq (u \circ t)(\pi)$ Right-hand side of OI

(7) $(t \circ u)(\pi) \sim (u \circ t)(\pi)$ From (3) and (6)   □

We shall now illustrate how this technical result can be applied to the paparazzi example. We use the following pair of transformative decision rules, and we assume without further ado that both rules satisfy the OI-condition.

The principle of insufficient reason (ir): *If π is a formal decision problem in which the probabilities of the states are unknown, then it may be transformed into a formal decision problem π′ in which equal probabilities are assigned to all states.*

Merger of states (ms): *If two or more states yield identical outcomes under all acts, then these repetitious states should be collapsed into one, and if the probabilities of the two states are known they should be added.*

It can now be easily verified that Table 2.15 can be obtained from 2.13 by first applying the ir rule and then the ms rule. Furthermore, 2.17 can be obtained from 2.13 by first applying the ms rule and then the ir rule. Because of Theorem 2.1 we know that (ir ∘ ms)(π) ∼ (ms ∘ ir)(π). Hence, anyone who thinks that the two transformative rules we use satisfy the OI-condition is logically committed to the view that the two formalisations, 2.15 and 2.17, are equally reasonable (and at least as good as 2.13). However, although equally reasonable, 2.15 and 2.17 are of course very different – the expected utility principle even recommends different acts!

Exercises

2.1 If you play roulette in Las Vegas and bet on a single number, the probability of winning is 1/38: There are 38 equally probable outcomes of the game, viz. 1–36, 0 and 00. If the ball lands on the number you have chosen the croupier will pay you 35 times the amount betted, and return the bet.

   (a) Formalise and visualise the decision problem in a decision matrix.
   (b) Formalise and visualise the decision problem in a decision tree.
   (c) How much money can you expect to lose, on average, for every dollar you bet?

2.2 Formalise the following decision problem, known as Pascal's wager:

   God either exists or He doesn't. …. It is abundantly fair to conceive, that there is at least 50% chance that the Christian Creator God does in fact exist. Therefore, since we stand to gain eternity, and thus infinity, the wise and safe choice is to live as though God does exist. If we are right, we gain everything, and lose nothing. If we are wrong, we gain nothing and lose nothing. Therefore, based on simple mathematics, only a fool would choose to live a Godless life. Since you must choose, let us see which interests you least. You have nothing to lose. Let us estimate these two chances. If you gain, you gain all; if you lose, you lose nothing. Wager, then, without hesitation that He is. (Pascal 1660: §233)

2.3 (a) Is Pascal's argument convincing? (b) Is it really necessary for Pascal to assume that, "there is at least 50% chance that the Christian Creator God does in fact exist"? What if the probability is much lower?

2.4 Congratulations! You have won a free holiday in a city of your choice: London, New Delhi or Tokyo. You have been to London before, and you know that the city is okay, but expensive. New Delhi would be very exciting, unless you get a stomach infection; then it would be terrible. Tokyo would be almost as exciting, given that it is not too cold; then the trip would be rather boring.

(a) Formalise and visualise the decision problem in a decision matrix.

(b) Formalise and visualise the decision problem in a decision tree.

(c) Represent the five possible outcomes in an ordinal scale.

2.5 A friend offers you to invest all your savings, $100,000, in his dot com company. You find it very hard to understand the business plan he presents to you, but your friend tells you that your $100,000 will 'certainly' be worth at least $10M within two years. Naturally, your friend may be right, but he may also be wrong – you feel that you cannot estimate the probabilities for this. Consider the decision matrix below. What is wrong with this formalisation?

	Friend is right	Friend is wrong
Invest	$10M	$0
Do not	$100,000	$100,000

2.6 Visualise the following vector (which is written on a single line, to save space) in a decision matrix: $[[a_1, a_2, a_3]; [s_1, s_2]; [(a_1, s_1)=p, (a_1, s_2)=q, (a_2, s_1)=r, (a_2, s_2)=s, (a_3, s_1)=t, (a_3, s_2)=u]]$.

2.7 Explain the difference between (a) ordinal and cardinal scales, and (b) interval scales and ratio scales.

2.8 Your rich aunt died some time ago. In her will she stipulated that you shall receive a painting of your choice from her collection of impressionist art. The aesthetical values of her four paintings are, as measured on your personal interval scale, as follows: Manet 5,000; Monet 8,000; Pissarro 6,000; Renoir 2,000. Which of the following scales can be obtained from the original scale by a positive linear transformation?

(a) Manet 8; Monet 11; Pissarro 9; Renoir 5
(b) Manet −250; Monet 2,750; Pissarro 750; Renoir −3,250
(c) Manet 1,000; Monet 3,000; Pissarro 2,950; Renoir 995

2.9 Show that scales (a)–(c) are equivalent ordinal scales.

2.10 Suppose that scale (a) in Exercise 2.8 is a ratio scale. Show that neither (b) nor (c) is equivalent to that ratio scale.

2.11 Prove that every ratio scale is an interval scale, and that every interval scale is an ordinal scale.

2.12 Suppose f is a ratio scale that can be transformed into f' by multiplying all values of f by k. Show that $f(x) = \frac{1}{k} \cdot f'(x)$.

Solutions

2.1 (a)

	Right number (1/38)	Wrong number (37/38)
Bet	$35	−$1
Do not	$0	$0

(b)

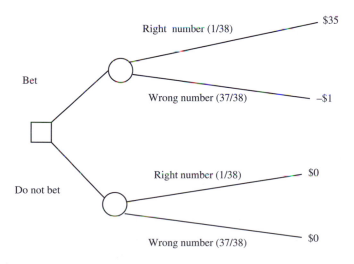

(c)

$$\left(\$(35+1) \cdot \frac{1}{38} + \$0 \cdot \frac{37}{38} \right) - \$1 \approx -\$.053$$

2.2

	God exists (1/2)	God does not exist (1/2)
Live as if God exists	Heaven – infinite gain.	You gain nothing and lose nothing.
Live a Godless life	Hell – infinite loss.	You gain nothing and lose nothing.

2.3 (a) Pascal's point is that the expected value of believing in God exceeds that of not believing. The secondary literature on this argument is massive, so I leave it to the reader to make up his or her own mind about it. (b) The assumption about a 50% probability is not important. All that matters is that the probability that God exists is > 0. (Since k times infinity equals infinity.)

2.4 We define four states: $s_1 =$ If you are in Delhi you get a stomach infection, and it is cold in Tokyo; $s_2 =$ If you are in Delhi you get a stomach infection, and it is not cold in Tokyo; $s_3 =$ If you are in Delhi you do not get a stomach infection, and it is cold in Tokyo; $s_4 =$ If you are in Delhi you do not get a stomach infection, and it is not cold in Tokyo. Now we draw the following matrix.

	s_1	s_2	s_3	s_4
London	Okay, but expensive	Okay, but expensive	Okay, but expensive	Okay, but expensive
New Delhi	Terrible!	Terrible!	Very exciting!	Very exciting!
Tokyo	Rather boring.	Almost as exciting as N.D.	Rather boring.	almost as exciting as N.D.

(b) Let the states be defined as in (a).

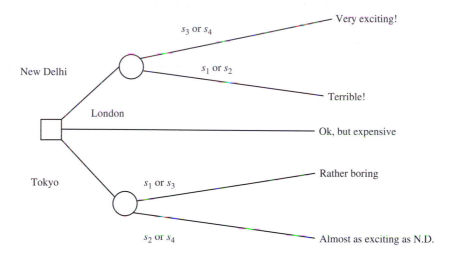

(c) Here is my suggestion: Very exciting! = 5, Almost as exciting as N.D. = 4, Okay, but expensive = 3, Rather boring = 2, Terrible = 1.

2.5 Your friend can be wrong in many ways. If you get $5M back your friend was literally speaking wrong, but you would presumably find that outcome quite pleasing. A satisfactory formalisation should take this into account, e.g. by listing each possible (monetary) outcome as a separate state, or by distinguishing between only two states: *satisfactory profit* and *unsatisfactory profit*.

2.6

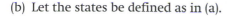

	s_1	s_2
a_1	p	q
a_2	r	s
a_3	t	u

2.7 Explicit answers can be found in the text.

2.8 (a) and (b)

2.9 For each scale f it holds that f(Monet) > f(Pissarro) > f(Manet > f(Renoir)

2.10 (a) and (b) cannot be equivalent, because $8/11 \neq -250/2{,}750$. Furthermore, (a) and (c) cannot be equivalent, because $8/11 \neq 1{,}000/3{,}000$.

2.11 This is trivial. Let $k = 0$ in (5), and note that (1) will hold true for all positive constants k and all constants m in (5).

2.12 This is also trivial. By definition, $f'(x) = k \cdot f(x)$; hence, $f(x) = \frac{1}{k} \cdot f'(x)$.

3 Decisions under ignorance

Jane is having a romantic dinner with her fiancé in a newly opened French bistro in Santa Barbara, California. After having enjoyed a vegetarian starter, Jane has to choose a main course. There are only two options on the menu, *Hamburger* and *Lotte de mer*. Jane recalls that *Lotte de mer* means monkfish, and she feels that this would be a nice option as long as it is cooked by a first-class chef. However, she has some vague suspicions that this may not be the case in this particular restaurant. The starter was rather poor and cooking monkfish is difficult. Virtually any restaurant can serve edible hamburgers, however.

Jane feels that she cannot assign any probability to the prospect of getting good monkfish. She simply knows too little about this newly opened restaurant. Because of this, she is in effect facing a decision under ignorance. In decision theory *ignorance* is a technical term with a very precise meaning. It refers to cases in which the decision maker (i) knows what her alternatives are and what outcomes they may result in, but (ii) she is unable to assign any probabilities to the states corresponding to the outcomes (see Table 3.1). Sometimes the term 'decision under uncertainty' is used synonymously.

Jane feels that ordering a hamburger would be a safe option, and a hamburger would also be better than having no main course at all. Furthermore, she feels that good monkfish is better than an edible hamburger, but terrible monkfish is worse than having no main course at all. This ranking of outcomes can be represented on an ordinal scale, as explained in Chapter 2. Since ordinal scales only preserve the order between objects, any set of numbers will do as long as better outcomes are represented by higher numbers. An example is given in Table 3.2.

Which alternative should Jane choose? To answer this question the decision maker has to apply some decision rule, or provide some other

Table 3.1

	Good chef	Bad chef
Monkfish (lotte de mer)	good monkfish	terrible monkfish
Hamburger	edible hamburger	edible hamburger
No main course	hungry	hungry

Table 3.2

	s_1	s_2
a_1	4	1
a_2	3	3
a_3	2	2

kind of structured reasoning for choosing one alternative over another. This chapter surveys some of the most influential rules for making decisions under ignorance.

3.1 Dominance

When Jane is about to choose a main course, rationality forbids her to choose a_3. (See Table 3.2.) This is because a_2 will lead to a better outcome than a_3 no matter which state happens to be the true state of the world; decision theorists say that alternative a_2 *dominates* alternative a_3. The widely accepted *dominance principle* prescribes that dominated acts must not be chosen.

Here is another example. Imagine that you wish to travel from London to Las Vegas. This route is serviced by two airlines, Black Jack airlines and Air Mojave, both of which offer equivalent fares and service levels. Just before you are about to book your ticket you read in *The Times* that Air Mojave is on the brink of bankruptcy. You therefore reason as follows: "No matter whether the information about Air Mojave is correct or not, I have nothing to lose by booking my ticket with Black Jack airlines. The fare and service is the same, but I will not face the same risk of losing my money. It is therefore more rational to book with Black Jack instead of Mojave." The decision matrix is given in Table 3.3.

You think that having a valid ticket is better than not having one. This ordinal relation can be represented by two arbitrary numbers, say 1 and 0.

Table 3.3

	Information correct	Information not correct
Black Jack airlines	Valid ticket	Valid ticket
Air Mojave	No valid ticket	Valid ticket

Table 3.4

	s_1	s_2
a_1	1	1
a_2	0	1

This yields the decision matrix in Table 3.4, in which it can be clearly seen that a_1 dominates a_2.

We are now in a position to state the dominance principle in a slightly more precise way. Briefly put, we wish to formalise the idea that one alternative is better than another if (but not only if) the agent knows for sure that she will be at least as well off if the first alternative is chosen, no matter what state of the world turns out to be the true state. Let $\succ$ be a relation on acts, such that $a_i \succ a_j$ if and only if it is more rational to perform act a_i rather than act a_j. This means that whenever $a_i \succ a_j$ is true, then a_i ought, rationally, to be chosen over a_j. Furthermore, $a_i \succeq a_j$ means that a_i is at least as rational as a_j, and $a_i \sim a_j$ means that the two acts are equally rational. Let v be an ordinal function that assigns values to outcomes, conceived of as ordered pairs of acts and states. For example, $v(a_1, s_1) = 1$ means that the value of doing a_1 in the case that s_1 is the true state of the world is 1.

The dominance principle comes in two versions; the Weak and the Strong dominance principle.

Weak Dominance: $a_i \succeq a_j$ if and only if $v(a_i, s) \geq v(a_j, s)$ for every state s.

Strong Dominance: $a_i \succ a_j$ if and only if $v(a_i, s_m) \geq v(a_j, s_m)$ for every state s_m, and there is some state s_n such that $v(a_i, s_n) > v(a_j, s_n)$.

Weak Dominance articulates the plausible idea that one alternative is at least as rational as another if all its outcomes, under every possible state of the world, will be at least as good as that of the other alternative. Strong Dominance stipulates that for an alternative to be more rational than another, two conditions must be fulfilled. First, its outcomes under all states must be at least as good as those of the other alternative. Secondly, there must also be *at least one* state under which the outcome is strictly better than that of the other alternative. For example, booking a ticket with Black Jack airlines strongly dominates booking a ticket with Air Mojave.

The dominance principle can also be applied to decisions under risk. If we know the probabilities of the states, it still makes sense to choose one alternative over another if it is certain to lead to at least as good an outcome no matter which state turns out to be the true state. Hence, it does not really matter if the probabilities of the states involved are known by the decision maker or not. All that matters is that we know that we will be at least as well off if we choose one alternative over another, no matter which state of the world turns out to be the true state.

Unfortunately, the dominance principle cannot tell the whole truth about decision making under ignorance. It merely provides a partial answer. This is because it sometimes fails to single out a set of optimal acts. For example, in Jane's decision at the restaurant in Santa Barbara, act a_1 is not dominated by a_2, nor is a_2 dominated by a_1. All that can be concluded is that it would *not* be rational to do a_3. This means that some additional criterion is needed to reach a decision.

3.2 Maximin and leximin

The maximin principle focuses on the worst possible outcome of each alternative. According to this principle, one should *maximise* the *minimal* value obtainable with each act. If the worst possible outcome of one alternative is better than that of another, then the former should be chosen. Consider the decision matrix in Table 3.5.

The worst possible outcome of each alternative is underlined. The best of these worst-case scenarios is both underlined and printed in bold

Table 3.5

	s_1	s_2	s_3	s_4
a_1	6	9	3	0
a_2	−5	7	4	12
a_3	6	4	5	2
a_4	14	−8	5	7

Table 3.6

	Good chef	Bad chef
Monkfish	Best outcome	Worst outcome
Hamburger	2nd best outcome	2nd best outcome
No main course	3rd best outcome	3rd best outcome

face. The maximin rule thus prescribes that one should choose alternative a_3, because the worst possible outcome of that act is 2, which is better than the worst-case scenarios of all other alternatives (0, −5, and −8).

The maximin rule is easy to formalise. Let $\min(a_i)$ be the minimal value obtainable with act a_i. Then,

Maximin: $a_i \succeq a_j$ if and only if $\min(a_i) \geq \min(a_j)$.

The maximin rule does not require that we compare outcomes on an interval scale. All we need is an ordinal scale, since the 'distance' between outcomes is irrelevant. In order to illustrate this point, we return to Jane in Santa Barbara. The matrix in Table 3.6 comprises enough information for applying the maximin rule.

Clearly, the maximin rule prescribes that Jane ought to order a hamburger, because the worst possible outcome would then be better than if any other alternative is chosen.

If the worst outcomes of two or more acts are equally good the maximin rule tells you to be indifferent. Hence, if the outcome of some act will be either $1 or $2, while some other act will give you either $10,000 or $1, you have to be indifferent between those options. This does not ring true. However, the standard remedy to this objection is to invoke the *lexical* maximin rule, which is slightly more sophisticated. It holds that if the worst outcomes are equal, one should choose an alternative such that the *second worst* outcome is certain to be as good as possible. If this does not single out a unique act, then the *third worst* outcome should be considered, and so on. Hence, unless all possible outcomes of the alternatives are exactly parallel, the lexical maximin rule (leximin rule) will at some point single out one or several acts that are better than all the others. Consider the example in Table 3.7, in which the traditional maximin rule ranks all alternatives as equally good, whereas the leximin rule recommends a_3.

Table 3.7

	s_1	s_2	s_3	s_4
a_1	6	14	8	5
a_2	5	6	100	100
a_3	7	7	7	5

Table 3.8

	s_1	s_2	s_3	s_4	s_5	s_6	s_7	s_8	s_9	s_{10}
a_1	1	100	100	100	100	100	100	100	100	100
a_2	1.1	1.1	1.1	1.1	1.1	1.1	1.1	1.1	1.1	1.1

To formalise the leximin rule, let $\min^1(a_i)$ be the value of the worst outcome of act a_i, and $\min^2(a_i)$ be the value of its second worst outcome, and $\min^n(a_i)$ be the value of the nth worst outcome of a_i. Then,

Leximin: $a_i \succ a_j$ if and only if there is some positive integer n such that $\min^n(a_i) > \min^n(a_j)$ and $\min^m(a_i) = \min^m(a_j)$ for all $m < n$.

What is the best argument for accepting maximin or leximin? It is tempting to reason as follows: In a certain sense, the maximin and leximin rules allow the decision maker to transform a decision under ignorance into a decision under certainty. Surely, a decision maker accepting the maximin or leximin rule does not know what the *actual* outcome of her choice will be. However, she knows for sure what the *worst possible* outcome will be. This is a kind of partial certainty, which many people seem to find attractive. If one accepts the premise that the worst possible outcome should guide one's decisions, then the maximin and leximin rules allows one to make decisions without fearing the actual outcome – the actual outcome will always be at least as good as the worst possible outcome.

However, in some decision problems it seems rather irrational to focus entirely on the worst-case scenario. Consider the example in Table 3.8.

The maximin and leximin rules prescribe that a_2 should be chosen over a_1. However, from a normative point of view this appears a bit strange. Most people would surely choose a_1, because most people would judge that it is worth sacrificing 0.1 units of value for the possibility of getting 100 units.

Table 3.9

	s_1	s_1 or s_2 or s_3 ... or s_{10}
a_1	1	100
a_2	1.1	1.1

What could one say in reply? To start with, note that the decision matrix in Table 3.8 gives the false impression that 100 units of value will be a much more *probable* outcome than 1 if the first alternative is chosen. However, the decision maker is making a decision under ignorance, so nothing can be concluded about the probabilities of the outcomes. The mere fact that there are more states corresponding to 100 rather than 1 does not imply that 100 is a more probable outcome than 1. In fact, by merging the nine rightmost states (s_2–s_{10}) into a single (disjunctive) state, we get Table 3.9, which is an equally good representation of the same decision problem.

Now, one no longer gets the false impression that 100 is a more probable outcome than 1. But despite this manoeuvre, it may still be reasonable to maintain that the first alternative is better than the second, contrary to the recommendations of the maximin and leximin rules. Most of us would probably agree that it is rational to trade the risk of getting 1 instead of 1.1 for a chance of getting 100.

3.3 Maximax and the optimism–pessimism rule

The maximin rule singles out the worst possible outcome of a decision as its only normatively relevant feature. However, one could of course equally well single out the *best* possible outcome as being particularly relevant. As I write this text, best-selling author J.K. Rowling is just about to release her seventh Harry Potter book. We all know that it will sell millions of copies. Clearly, as Ms Rowling wrote her first story about an unknown boy called Harry Potter and submitted it to a publisher, she did not think much about the worst-case scenario, i.e. to have the manuscript rejected. What triggered her to actually submit the book to the publisher was no doubt her desire to achieve the *best* possible outcome of this decision. Luckily for her, the best possible outcome also turned out to be the *actual* outcome. If all authors considering submitting a manuscript always decided to focus on the worst-case scenario, i.e. rejection and disappointment, no books would ever be published.

The J.K. Rowling example serves as a nice illustration of the maximax rule. Advocates of the maximax rule maintain that rationality requires us to prefer alternatives in which the best possible outcome is as good as possible, i.e. one should *maximise* the *maximal* value obtainable with an act. The maximax rule has surprisingly few adherents among decision theorists, but as far as I can tell, it is about as reasonable as the maximin rule. However, maximax also serves as a helpful introduction to another rule, which has attracted considerably more attention by decision theorists. This is the optimism–pessimism rule, originally proposed by Hurwicz. (A synonymous name for this rule is the alpha-index rule.) The optimism–pessimism rule asks the decision maker to consider both the best and the worst possible outcome of each alternative, and then choose an alternative according to her degree of optimism or pessimism. In order to render this suggestion precise, suppose that a decision maker's degree of optimism can be represented by a real number α between 0 and 1, such that $\alpha = 1$ corresponds to maximal optimism and $\alpha = 0$ to maximal pessimism. Then, if $\max(a_i)$ is the best possible outcome of alternative a_i and $\min(a_i)$ its worst possible outcome, its value is $\alpha \cdot \max(a_i) + (1 - \alpha) \cdot \min(a_i)$. Naturally, α is assumed to be fixed throughout the evaluation of all alternatives a_i.

Consider the example in Table 3.10. Note that we have omitted the states corresponding to the outcomes. It is not necessary to always write them down.

The best possible outcome of a_1 is 100 and the worst possible outcome is 10. That is, $\max(a_1) = 100$ and $\min(a_1) = 10$. Hence, if the decision maker is optimistic to degree 0.7, it follows that the total value of a_1 is $0.7 \cdot 100 + (1 - 0.7) \cdot 10 = 73$. Furthermore, the best possible outcome of alternative a_2 is 90, whereas its worst outcome is 50. Hence, the value of this alternative is $0.7 \cdot 90 + (1 - 0.7) \cdot 50 = 78$. Since $78 > 73$, it follows that alternative a_2 is better than alternative a_1.

Table 3.10

a_1	55	18	28	10	36	100
a_2	50	87	55	90	75	70

Optimism–Pessimism: $a_i \succ a_j$ if and only if $\alpha \cdot \max(a_i) + (1-\alpha) \cdot \min(a_i) > \alpha \cdot \max(a_j) + (1-\alpha) \cdot \min(a_j)$

Note that if $\alpha = 0$, then the optimism–pessimism rule collapses into the maximin rule, since $0 \cdot \max(a_i) + (1 - 0) \cdot \min(a_i) = \min(a_i)$. Furthermore, the optimism–pessimism rule will be identical with the maximax rule if $\alpha = 1$. The optimism–pessimism rule can therefore be conceived of as a generalisation of the maximin and maximax rules.

The optimism–pessimism rule clearly requires that value can be measured on an interval scale, because if one were to measure value on an ordinal scale it would make no sense to perform arithmetical operations such as multiplication and addition on the values. However, the maximin and maximax rules only require that value is measured on ordinal scales. The claim that the optimism–pessimism rule is a generalisation of maximin and maximax must therefore be taken with a pinch of salt.

The main advantage of the optimism–pessimism rule is that it allows the decision maker to pay attention to the best-case *and* the worst-case scenarios at the same time. The α-index describes the relative importance of both outcomes. The obvious objection is, of course, that it seems irrational to totally ignore the outcomes in between. Should one really ignore all the non-minimal and non-maximal outcomes? Please take a look at Table 3.11. The max and min values of both alternatives are the same, so for every α it holds that both alternatives are equally good. However, intuitively it seems reasonable to choose the first alternative, because under almost all states it will lead to a significantly better outcome, and there is only one state in which it leads to a slightly worse outcome.

An obvious objection to this argument is that it indirectly invokes a sort of probabilistic reasoning. Instead of assigning numerical probabilities to the states, the argument falsely presupposes that one alternative is more likely than another if there are more states leading to a certain outcome. In a decision under ignorance nothing is known about probabilities.

Table 3.11

a_1	10	10	99	99	...	99	99	100
a_2	10	11	11	11	...	11	11	100

However, the basic objection remains. Why is it rational to focus exclusively on the best and the worst possible outcomes? What about the other, non-extreme outcomes? A natural response might be to assign α-values to all outcomes, not only to the best and worst ones. If there are n possible outcomes, ordered from the best to the worst, the numbers $\alpha_1, \alpha_2, \ldots, \alpha_n$ should be chosen such that $\alpha_1 + \alpha_2 + \cdots + \alpha_n = 1$. Each number α_i describes the relative weight of outcome i. The value of the act is thereafter calculated as above, i.e. by multiplying the α-index and value of each outcome and thereafter adding all α-weighted outcomes. It should be emphasised that $\alpha_1, \alpha_2, \ldots, \alpha_n$ cannot be interpreted as probabilities, because $\alpha_1, \alpha_2, \ldots, \alpha_n$ are chosen according to how important the decision maker feels that the best, and the second-best, and the third-best outcomes are, and so on. There is in general no reason to think that the relative importance of an outcome corresponds to its probability.

3.4 Minimax regret

Imagine that you are invited to invest in a newly launched software company. Some of your friends, who invested in similar companies a few years ago, made a fortune. Of course, you may lose the money, but the investment is rather small, so the fact that the company may go bankrupt is not a genuine issue of concern. However, you know that if you *refrain* from making the investment and the company turns out to be yet another success story, you will feel a lot of regret. Some academics find this intuition important and argue that the concept of regret is relevant to rational decision making. Briefly put, advocates of the *minimax regret* rule maintain that the best alternative is one that minimises the maximum amount of regret. An example is given in Table 3.12.

We assume that outcomes are compared on an interval scale, not merely an ordinal scale. The numbers in bold denote the best alternative relative

Table 3.12

a_1	12	8	20	**20**
a_2	10	**15**	16	8
a_3	**30**	6	25	14
a_4	20	4	**30**	10

Table 3.13

a_1	**−18**	−7	−10	0
a_2	**−20**	0	−14	−12
a_3	0	**−9**	−5	−6
a_4	−10	**−11**	0	−10

to each state. For example, 20 is highlighted in the rightmost column because a_1 would be the best alternative to choose given that the rightmost state was to occur. Now, for each outcome in the decision matrix, regret values can be calculated by subtracting the value of the best outcome of each state (the number in bold face) from the value of the outcome in question. For example, the regret value of the lowermost outcome in the rightmost column is −10, because 10 − 20 = −10, and the regret value of the second lowermost outcome in the rightmost column is −6, because 14 − 20 = −6. By doing the analogous calculations for all outcomes in the decision matrix, we obtain a *regret matrix*. In the example we are considering, this is given in Table 3.13.

In the regret matrix, the numbers in bold denote the maximum regret value of each alternative act. For example, −18 is worse than −7, −10 and 0. Hence, −18 is the highest amount of regret obtainable with a_1. Note that all regret values will always be negative or zero. There is no such thing as 'positive' regret.

The minimax regret rule urges decision makers to choose an alternative under which the maximum regret value is as low as possible. Since −9 corresponds to less regret than −18, −20 and −11, act a_3 is the best alternative in this example. Note that once the regret matrix has been obtained from the original decision matrix, the minimax regret rule is parallel to the maximin rule. In a certain sense, both rules select an alternative under which the worst-case scenario is as good as possible, but the two rules operate on different descriptions of the relevant worst-case scenario (actual outcome versus possible regret). Another important difference is that the minimax regret rule requires that outcomes are compared on a cardinal scale, whereas the maximin rule operates on ordinal values.

Unsurprisingly, the minimax regret rule will look a bit complex when formalised. Let Max{ } be a function that returns the maximum value of a

Table 3.14

a_1	12	8	20	20
a_2	10	**15**	16	8
u_3	**30**	6	25	14
a_4	20	4	**30**	10
a_5	−10	10	10	**39**

Table 3.15

a_1	−18	−7	−10	**−19**
a_2	−20	0	−14	**−31**
a_2	0	−9	−5	**−25**
a_3	−10	−11	0	**−29**
a_5	**−40**	−5	−20	0

sequence of real numbers contained in a set { }. Furthermore, let max(s_i) be a function that returns the best possible outcome that may be achieved with any alternative if s_i is the true state of the world. Then, the minimax regret rule can be written as follows.

Minimax regret: $a_1 \succ a_2$ if and only if max{($v(a_1, s_1)$ − max(s_1)), ($v(a_1, s_2)$ − max(s_2)), ...} > max{($v(a_2, s_1)$ − max(s_1)), ($v(a_2, s_2)$ − max(s_2)), ...}

The most prominent objection to minimax regret is the objection from *irrelevant alternatives*. Consider the decision matrix in Table 3.14, which is obtained from Table 3.12 by adding a new alternative, a_5.

The regret matrix corresponding to Table 3.14 is given in Table 3.15.

It can be easily verified that in this matrix, alternative a_1 comes out as the best alternative according to the minimax regret rule – not a_3 as in the previous example. Why is this problematic? The answer is that the ranking of the alternatives has been altered *by adding a non-optimal alternative*. That is, by simply adding a bad alternative that will not be chosen (a_5), we have altered the order between other, better alternatives. Intuitively, this is strange, because why should one's choice between, say, a_1 and a_3 depend on whether a suboptimal alternative such as a_5 is considered or not? It seems reasonable to maintain that a normatively plausible decision rule

---

Box 3.1   Minimax and the objection from irrelevant alternatives

The objection from irrelevant alternatives is often thought to be a decisive argument against the minimax regret rule. However, it is far from clear that the objection is as strong as it may first appear to be: Imagine that Bizarro, the world famous artist, produced fifty paintings in his lifetime. Some of these masterpieces are predominately red while the others are predominately blue. For many years no more than two paintings, one red and one blue, have been in private ownership. All remaining paintings are government property and stored in buildings which are closed to the public. However, as luck has it you are rather well off and can afford to buy any one of the two paintings in private ownership. Upon studying both paintings you decide that you much prefer the red to the blue. In an attempt to reduce its budget deficit, the government suddenly decides to sell off one of its Bizarros. The painting that now becomes available for choice is a red one. You find the new painting quite unappealing, and you like both paintings in the initial choice set more than the new one. Furthermore, for some reason, never mind why, the addition of the new painting to the choice set does not affect the prices of the old ones. In fact, *everything* else remains the same. The only difference is that one more painting has been added to the choice set. Nonetheless, you no longer prefer the first red Bizarro to the blue. Since the addition of a second red painting to the choice set makes the blue *unique* (it is the only blue Bizarro in the choice set), you switch your preference. You now prefer the blue over the red, and it is the mere addition of a second red painting to the choice set that makes you switch your preference.

It can hardly be denied that people sometimes reason like this. But is this holistic reasoning rational? I think so. When stating preferences over a set of alternatives, it is rational to compare each alternative with the entire set of alternatives, even if some of those alternatives are certain not to be chosen. By adding a new object to the choice set, the relational properties of the initial objects may change, and sometimes our preferences are based on relational properties, such as uniqueness.

---

must not be sensitive to the addition of irrelevant alternatives. Therefore, since the minimax regret rule does not meet this requirement it must be ruled out as an illegitimate rule. However, as pointed out in Box 3.1 this objection is not undisputed.

3.5 The principle of insufficient reason

A number of prominent thinkers, including Bernoulli (1654–1705) and Laplace (1749–1827), have defended a rule known as the *principle of insufficient reason*. Imagine that you have booked a weekend trip to a foreign country. Unfortunately, you know nothing about the weather at your destination. What kind of clothes should you take? In your wardrobe you have clothes for three types of weather, viz. hot, medium and cold weather. The decision matrix is given in Table 3.16.

The principle of insufficient reason prescribes that if one has *no* reason to think that one state of the world is more probable than another, then all states should be assigned *equal* probability. In the example above there are three possible states, so the probability assigned to each state will be 1/3 (Table 3.17). More generally speaking, if there are n states the probability $p(s)$ assigned to each state will be $1/n$.

By applying the principle of insufficient reason, an initial decision problem under ignorance is transformed into a decision problem under risk, that is, a decision problem with known probabilities and values. As will be further explained in Chapter 4, many decision theorists think

Table 3.16

	Actual temperature is 100°F	Actual temperature is 50°F	Actual temperature is 0°F
Bring clothes suitable for 100°F	15	0	−30
Bring clothes suitable for 50°F	0	15	0
Bring clothes suitable for 0°F	−15	0	15

Table 3.17

	s_1 (1/3)	s_2 (1/3)	s_3 (1/3)
a_1	15	0	−30
a_2	0	15	0
a_3	−15	0	15

that one should maximise expected value in decisions under risk. In our example, the expected value of bringing clothes suitable for 100°F is $1/3 \cdot 15 + 1/3 \cdot 0 + 1/3 \cdot -30 = -5$. The expected value of the second alternative, i.e. bringing clothes suitable for 50°F, is $1/3 \cdot 0 + 1/3 \cdot 15 + 1/3 \cdot 0 = 5$, and the expected value of the third alternative is $1/3 \cdot -15 + 1/3 \cdot 0 + 1/3 \cdot 15 = 0$. Hence, it is rational to choose the second alternative, i.e. to bring clothes suitable for 50°F.

Note that the principle of insufficient reason in itself does not recommend any particular act. It merely *transforms* a decision under ignorance into a decision under risk, by assigning equal probabilities to all states. Hence, the principle of insufficient reason could be combined with any decision rule for making decisions under risk. However, nearly everyone advocating the principle of insufficient reason also thinks that decisions under risk should be taken by calculating expected value. This composite version of the principle of insufficient reason can be formalised in the following way.

The Principle of Insufficient Reason: $a_i \succ a_j$ if and only if $\sum_{x=1}^{n} \frac{1}{n} v(a_i, s_x) > \sum_{x=1}^{n} \frac{1}{n} v(a_j, s_x)$

An obvious problem with the principle of insufficient reason is that it is extremely sensitive to how states are individuated. Suppose that before embarking on a trip you consider whether to bring an umbrella or not. As before, you know nothing about the weather at your destination. If the formalisation of the decision problem is taken to include only two states, viz. rain and no rain, the probability of each state will be 1/2. However, it seems that one might just as well go for a formalisation that divides the space of possibilities into three states, viz. heavy rain, moderate rain and no rain. If the principle of insufficient reason is applied to the latter set of states, their probabilities will be 1/3. In some cases this difference will affect our decisions. Hence, it seems that anyone advocating the principle of insufficient reason must ascertain the rather implausible hypothesis that there is only one correct way of making up the set of states.

Traditionally, decision theorists advocating the principle of insufficient reason have tried to make it more plausible by stressing the importance of considering *symmetric* states. For example, if someone gives you a die with n sides and you have no reason to think that any side is more likely than another (i.e. you have no reason to think that the die is biased), then you should assign a probability of $1/n$ to each side. However, it seems that not all events can be described in symmetric terms, at least not in a way that justifies the conclusion that they are equally probable. Whether Ann's marriage will be a happy one depends on her future emotional attitude towards her husband. According to one description, she could be either in love or not in love with him; then the probability of both states would be $1/2$. According to another equally plausible description, she could either be deeply in love, a little bit in love or not at all in love with her husband; then the probability of each state would be $1/3$. In both cases the symmetry condition seems to be satisfied, which shows that it is too weak.

Here is another objection to the principle of insufficient reason: If one has *no* reason to think that one state is more probable than another, it seems strange to conclude that they are *equally* probable. If one has *no* reason to think that some state is *twice* as probable as another, we would not conclude that it is twice as probable as the other. More generally speaking, the problem is that it seems *completely arbitrary* to infer that all states are equally probable. Any other distribution of probabilities seems to be equally justified (that is, not at all).

It might be replied that the claim that all states are equally probable is in a certain sense special. Because if one has no reason to think that s_1 is twice as probable as s_2, it may also hold true that one has no reason to think that s_2 is twice as probable as s_1. However, it cannot be the case that s_1 *is* twice as probable as s_2 and that s_2 *is* twice as probable as s_1. Only if we assign equal probabilities to all states can we avoid ascribing several *different* probabilities to one and the same state.

Even if this line of reasoning might have something going for it the basic objection remains. The principle of insufficient reason could equally well be taken to read, "If you have no reason for inferring anything about the probabilities of the states, then assign equal probabilities to all states." Once the principle is formulated in this way, it becomes natural to ask: Why favour any particular way of assigning probabilities to states over another? *Any* assignment of probability seems to be exactly as justifiable as another (that is, not at all). In the case with two states, $1/2$ and $1/2$ is just one of many options. One could equally well assign probability 0.6 to the first state and 0.4 to the second, and so on.

3.6 Randomised acts

Instead of selecting one act over another, one may instead toss a coin and, for example, choose the first alternative if it lands heads up and the second otherwise. Consider the decision problem in Table 3.18.

Randomising between a_1 and a_2 means that a new alternative is introduced, which is the act of randomising between the other acts, thereby transforming the initial decision matrix into a new matrix with three alternatives. If the randomisation is performed by tossing a fair coin the new decision matrix will be as in Table 3.19.

In the initial, non-randomised matrix, all acts come out equally well according to all decision rules discussed in this chapter. However, if we add the randomised act a_3, this act will be chosen by the maximin, leximin and minimax regret rules (as well as by the optimism–pessimism rule if $\alpha > 1/2$). Hence, a_3 is strictly better than both other acts. So it seems that something can be gained by allowing randomised acts.

It should be emphasised that the randomised act a_3 will not lead to a *certain* outcome worth 1/2, however. The number 1/2 is the *expected* value of randomising between a_1 and a_2 with a probability of 1/2. (Of course, any probability could be chosen; 1/2 is just an example.) This means that there are two cases to consider. To start with, consider the case in which the decision maker faces some decision problem only once. Suppose that you have been walking through the desert for days with no food or water. You now come to a crossroads. Should you choose path a_1 or a_2? If you make the right decision you will reach an oasis; otherwise you will die. What

Table 3.18

a_1	1	0
a_2	0	1

Table 3.19

a_1	1	0
a_2	0	1
a_3	1/2	1/2

would be the point of randomising between the two alternatives in this case? Depending on which path you choose you will either die or survive. To randomise between the two paths does not mean that you will 'die to degree 1/2', or get any other outcome between the two extremes. You will either die or reach the oasis. Hence, it is far from clear exactly what is gained by randomising in this type of case.

In the second type of case, you know that you will face the same type of decision over and over again. Therefore, if you randomise between a_1 and a_2, you will *on average* get 1/2. This conclusion is a bit strange, however. If one faces exactly the same decision problem over and over again, literally speaking, then the truth about the two states would be fixed once and for all. Hence, it would be better to choose either a_1 or a_2 the first time and then *adjust* one's future behaviour to the outcome of the first decision. If one chooses a_2 and gets 0, one should switch to a_1, which would then be certain to lead to 1 in all remaining decisions. Furthermore, if the truth about the states is *not* fixed once and for all, one should of course choose either a_1 or a_2, stick to it, and then adjust one's behaviour as one gradually learns more about the relative frequencies of the states.

Box 3.2   Axiomatic analysis of decisions under ignorance

The decision rules discussed in this chapter sometimes yield conflicting prescriptions, so not all of them can be accepted by a rational decision maker. For example, in Table 3.20 maximin recommends a_1, minimax regret and the principle of insufficient reason recommend a_2, whereas the optimism–pessimism rule recommends a_3 (for $\alpha = 1/2$).

Which rule is best? To argue that one rule is better than another because its normative recommendations are more intuitively plausible is unlikely to impress the decision theorist. Some stronger argument is needed. Traditionally, decision theorists seek to show that one rule is better than another by providing an axiomatic justification. Briefly put, the idea is to formulate a set of fundamental principles from which a decision rule can

Table 3.20

a_1	20	15	1
a_2	19	20	0
a_3	0	30	0

be derived – or, in case one wants to argue against a rule, to find a set of plausible conditions which are inconsistent with the rule in question. Several authors have suggested such axiomatic analyses, but the most accessible and elegant one was presented in a classic paper by Milnor.

Milnor assumed that a plausible decision rule must not only single out a set of optimal acts, but also provide an ordinal ranking of all alternatives. Let $\succeq$ be a binary relation on acts such that $a_i \succeq a_j$ if and only if it is at least as rational to choose a_i as a_j. Furthermore, $a_i \succ a_j$ means that $a_i \succeq a_j$ & not-($a_j \succeq a_i$), and $a_i \sim a_j$ means that $a_i \succeq a_j$ and $a_j \succeq a_i$. Milnor then asked us to consider a set of ten axioms. (For simplicity, we state the axioms informally.)

1. **Ordering**: $\succeq$ is transitive and complete. (See Chapter 5.)
2. **Symmetry**: The ordering imposed by $\succeq$ is independent of the labelling of acts and states, so any two rows or columns in the decision matrix could be swapped.
3. **Strict Dominance**: If the outcome of one act is strictly better than the outcome of another under every state, then the former act is ranked above the latter.
4. **Continuity**: If one act weakly dominates another in a sequence of decision problems under ignorance, then this holds true also in the limit decision problem under ignorance.
5. **Interval scale**: The ordering imposed by $\succeq$ remains unaffected by a positive linear transformation of the values assigned to outcomes.
6. **Irrelevant alternatives**: The ordering between old alternatives does not change if new alternatives are added to the decision problem.
7. **Column linearity**: The ordering imposed by $\succeq$ does not change if a constant is added to a column.
8. **Column duplication**: The ordering imposed by $\succeq$ does not change if an identical state (column) is added.
9. **Randomisation**: If two acts are equally valuable, then every randomisation between the two acts is also equally valuable.
10. **Special row adjunction**: Adding a weakly dominated act does not change the ordering of old acts.

How do these axioms relate to the rules discussed in this chapter? Consider Table 3.21. The symbol – means that a decision rule and axiom are incompatible, whereas × indicates that they are compatible. Furthermore, a decision rule is characterised by the axioms marked with

Table 3.21

	Maximin	Optimism– pessimism	Minimax regret	Insufficient reason
1. Ordering	⊗	⊗	⊗	⊗
2. Symmetry	⊗	⊗	⊗	⊗
3. Strict dominance	⊗	⊗	⊗	⊗
4. Continuity	⊗	⊗	⊗	⊗
5. Interval scale	×	⊗	×	×
6. Irrelevant alternatives	⊗	⊗	–	⊗
7. Column linearity	–	–	⊗	⊗
8. Column duplication	⊗	⊗	⊗	–
9. Randomisation	⊗	–	⊗	×
10. Special row adjunction	×	×	⊗	×

Table 3.22

a_1	1	2	a_1	11	2
a_2	0	3	a_2	10	3

the symbol ⊗, that is, those axioms constitute necessary and sufficient conditions for the decision rule in question.

The table summarises 44 separate theorems. We shall not prove all of them here, but for illustrative purposes we prove one theorem of each type. Consider the following examples.

Theorem 3.1   Column linearity is incompatible with maximin.

Proof   We construct a counter example to the claim that column linearity and maximin are compatible. Consider the pair of decision problems in Table 3.22.

The rightmost matrix can be obtained from the leftmost by adding 10 to its first column. However, in the leftmost matrix, the maximin rule prescribes a_1, whereas it prescribes a_2 in the rightmost one. Hence, contrary to what is required by column linearity, maximin entails that it is false that the ordering of acts remains unchanged if a constant is added to a column. □

Theorem 3.2   Randomisation is compatible with the principle of insufficient reason.

Proof According to the principle of insufficient reason, the value of an arbitrary act a_i equals $\sum_{m=1}^{n} \frac{1}{n} v(a_i, s_m)$. Call this sum z. Now, it follows that if two acts are to be equally good, then their values z and z' must be the same. Therefore, if we randomise between two acts with identical values z, then the value of the randomised act will also be z.   □

Theorem 3.3 The principle of insufficient reason is characterised by Axioms 1, 2, 3, 6 and 7.

Proof We wish to prove that Axioms 1, 2, 3, 6 and 7 are necessary and sufficient conditions for the principle of insufficient reason. I leave it to the reader to verify that they are necessary, i.e. that the principle of insufficient reason entails the axioms.

In order to prove that the axioms are sufficient we first prove the following lemma: The axioms guarantee that two acts which differ only in the order of their outcomes are ranked as equally good. In order to see why this is true, take two arbitrary acts and add a sequence of intermediate acts (by applying Axiom 6) such that any two acts differ only by a permutation of two outcomes. Our lemma now follows by applying Axiom 2 to each pair of successive rows.

Next, we wish to show that if the average outcomes of two acts a_i and a_j are the same, then the axioms guarantee that $a_i \sim a_j$. It is helpful to proceed in two steps. (i) We first use our lemma to permute the outcomes of a_i and a_j in order of increasing value, i.e. such that the worst outcome is listed first and the best last. (ii) In the second step we apply Axiom 7 and subtract from each column of a_i and a_j the number corresponding to its worst outcome. By repeating steps (i) and (ii) a finite number of times, we obtain a decision matrix with two rows in which all numbers are zero. (This is because we stipulated that the two acts have the same average outcome.) Hence, since the new versions have exactly the same outcomes, Axioms 1 and 3 imply that $a_i \sim a_j$.   □

Exercises

3.1 Which act(s) should be chosen in the decision problem below according to the following decision rule:
   (a) the maximin rule
   (b) maximax

(c) minimax regret

(d) the optimism–pessimism rule (for $\alpha > 1/4$)

(e) the principle of insufficient reason.

a_1	20	20	0	10
a_2	10	10	10	10
a_3	0	40	0	0
a_4	10	30	0	0

3.2 What is the main advantage of leximin as compared to maximin?

3.3 Construct a decision matrix that corresponds to the following regret table.

a_1	−5	−45	0
a_2	0	−30	−20
a_3	−10	0	−25

3.4 According to the minimax regret principle, regret is defined as the difference between the best outcome obtainable under a given state and the outcome obtained under that state with the chosen act. Now consider an alternative version of this principle, according to which regret is defined (relative to each act) as the difference between the best outcome and the worst outcome obtainable with the act in question. This modification of the minimax regret rule would give us a highly implausible decision rule. Why?

3.5 Prove that Axiom 8 (Column duplication) is incompatible with the principle of insufficient reason.

3.6 Prove that Axiom 7 (Column linearity) is incompatible with the optimism–pessimism rule.

3.7 Explain informally why Axiom 5 (Interval scale) is compatible with all of the following rules: (a) maximin, (b) optimism–pessimism, (c) minimax regret, (d) the principle of insufficient reason.

3.8 Rawls famously argued that the maximin rule could be used for formulating a theory of social justice. Very briefly put, he argued that a distribution A of social goods is better than an alternative distribution B if the worst-off person is better off in A than in B. Suppose that Rawls' theory of social justice is based on the optimism–pessimism rule instead of maximin. What would be the implications for social justice? (Assume that $1 > \alpha > 0$.)

Solutions

3.1 (a) a_2, (b) a_3, (c) a_4 (d) a_3, (e) a_1

3.2 An explicit answer is given in the text.

3.3 There are many solutions to this problem; here is one.

a_1	5	0	25
a_2	10	15	5
a_3	0	45	0

3.4 According to the modified rule an act that gives you, say, either 1 or 0 would be ranked higher than an act that yields either 100 or 98. This can't be right.

3.5 Here is a counter example: The matrix to the right can be obtained from the left by duplicating the second column. However, in the leftmost matrix the principle of insufficient reason is indifferent between a_1 and a_2, whereas in the rightmost matrix it recommends a_2.

a_1	1	0	a_1	1	0	1
a_2	0	1	a_2	0	1	1

3.6 Here is a counter example: The matrix to the right can be obtained from the left by adding 1 to the first column. However, in the leftmost matrix the optimism–pessimism rule is indifferent between a_1 and a_2 (for all α), whereas in the rightmost matrix it recommends a_1 if $\alpha > 1/2$, and a_2 if $\alpha < 1/2$ and is indifferent between them if and only if $\alpha = 1/2$.

a_1	1	0	a_1	2	0
a_2	0	1	a_2	1	1

3.7 None of these rules require any more information than what is contained in an interval scale. In particular, none of the rules require that we calculate *ratios* between outcomes.

3.8 The optimism–pessimism rule only considers the worst and the best possible outcomes of an act. All intermediate outcomes are irrelevant. If we translate this to a social context, in which each possible outcome corresponds to a single individual, we find that it does not matter if

99 out 100 people are badly off and 1 person is well off, or if 1 person is well off and 99 persons are badly off. Furthermore, since the best *possible* outcomes matter, it follows that we could make society better by making things much better for the best-off individual. I leave it to the reader to judge whether a theory of social justice thus modified is an improvement over Rawls' original theory.

4   Decisions under risk

Decisions under risk differ from decisions under ignorance in that the decision maker knows the probabilities of the outcomes. If you play roulette in Las Vegas you are making a decision under risk, since you then know the probability of winning and thus how much you should expect to lose. However, that you *know* the probability need not mean that you are in a position to immediately determine the probability or expected outcome. It is sufficient that you have enough information for figuring out the answer after having performed a series of calculations, which may be very complex. In this process your 'tacit knowledge' is made explicit. When you play roulette in Las Vegas and bet on a single number, the probability of winning is 1/38: There are 38 equally probable outcomes of the game, viz. 1–36 and 0 and 00, and if the ball lands on the number you have betted on the croupier will pay you 35 times the amount betted, and return your bet. Hence, if you bet $1, the expected payout is $(35+1) \cdot \dfrac{1}{38} + \$0 \cdot \dfrac{37}{38} = \$\dfrac{36}{38} \approx \$0.947$. This means that you can expect to lose about $1 − $0.947 = $.053 for every dollar betted.

According to the principle of maximising expected *monetary value* it would obviously be a mistake to play roulette in Las Vegas. However, this does not show that it is *irrational* to play roulette there, all things considered. First, the expected monetary value need not correspond to the overall value of a gamble. Perhaps you are very poor and desperately need to buy some medicine that costs $35. Then it would make sense to play roulette with your last dollar, since that would entitle you to a chance of winning just enough to buy the medicine. Second, it also seems clear that many people enjoy the sensation of excitement caused by betting. To pay for this is not irrational. (It is just vulgar!) Finally, one may also question the principle of maximising expected value as a general guide to risky choices. Is this really the correct way of evaluating risky acts?

In what follows we shall focus on the last question, i.e. we shall discuss whether it makes sense to think that the principle of maximising expected value is a reasonable decision rule to use in decisions under risk. Somewhat surprisingly, nearly all decision theorists agree that this is the case. There are no serious contenders. This is thus a significant difference compared to decision making under ignorance. As explained in Chapter 3, there is virtually no agreement on how to make decisions under ignorance.

That said, there is significant disagreement among decision theorists about how to *articulate* the principle of maximising expected value. The main idea is simple, but substantial disagreement remains about how to define central concepts such as 'value' and 'probability', and how to account for the causal mechanisms producing the outcomes. In this chapter we shall take a preliminary look at some aspects of these controversies, but it will take several chapters before we have acquired a comprehensive understanding of the debate.

It is worth noting that many situations outside the casino, i.e. in the 'real' world, also involve decision making under risk. For example, if you suffer from a heart disease and your doctor offers you a transplant giving you a 60% chance of survival, you are facing a decision under risk in which it seems utterly important to get the theoretical aspects of the decision right. It would thus be a mistake to think that decision making under risk is essentially linked to gambling. Gambling examples just happen to be a convenient way of illustrating some of the major ideas and arguments.

4.1 Maximising what?

The principles of maximising *expected monetary value* must not be confused with the principle of maximising *expected value*. Money is not all that matters, at least not to all of us. However, in addition to this distinction, we shall also introduce a new distinction between the principle of maximising *expected value* and the principle of maximising *expected utility*. The latter is a more precise version of the former, in which the notion of value is more clearly specified. This gives us three closely related principles, all of which are frequently mentioned in the literature.

1. The principle of maximising expected monetary value
2. The principle of maximising expected value
3. The principle of maximising expected utility

It is helpful to illustrate the difference between these principles in an example. Imagine that you are offered a choice between receiving a million dollars for sure, and receiving a lottery ticket that entitles you to a fifty per cent chance of winning either three million dollars or nothing (Table 4.1).

The expected monetary value (EMV) of these lotteries can be computed by applying the following general formula, in which p_1 is the probability of the first state and m_1 the monetary value of the corresponding outcome:

$$EMV = p_1 \cdot m_1 + p_2 \cdot m_2 + \cdots + p_n \cdot m_n \tag{1}$$

By applying (1) to our example, we find that the expected monetary values of the two lotteries are:

$$EMV(\text{Lottery A}) = \frac{1}{2} \cdot \$1M + \frac{1}{2} \cdot \$1M = \$1M.$$

$$EMV(\text{Lottery B}) = \frac{1}{2} \cdot \$3M + \frac{1}{2} \cdot \$0 = \$1.5M$$

However, even though EMV(Lottery B) > EMV(Lottery A), many of us would prefer a million for sure. The explanation is that the overall value to us of $3M is just slightly higher than that of $1M, whereas the value of $1M is much higher than the value of $0. Economists say that the *marginal value* of money is decreasing. The graph in Figure 4.1 describes a hypothetical relationship between money and value, for a poor person playing the National Lottery.

Note that the graph slopes upwards but with decreasing speed. This means that winning more is always better than winning less. However, the more one wins, the lower is the value of winning yet another million. That said, it is of course not a universal truth that the marginal value of

Table 4.1

	1/2	1/2
Lottery A	$1M	$1M
Lottery B	$3M	$0

Figure 4.1

money is decreasing. For people with very expensive habits the marginal value may be increasing, and it is even conceivable that it is negative for some people. Imagine, for instance, that you are a multi-billionaire. If you accidentally acquire another billion this will perhaps decrease your well-being, since more money makes it more likely that you will get kidnapped and you cannot stand the thought of being locked up in a dirty basement and having your fingers cut off one by one. If one is so rich that one starts to fear kidnappers, it would perhaps be better to get rid of some of the money.

Clearly, the principle of maximising expected value makes more sense from a normative point of view than the principle of maximising expected monetary value. The former is obtained from the latter by replacing m for v in the formula above, where v denotes value rather than money.

$$EV = p_1 \cdot v_1 + p_2 \cdot v_2 + \cdots + p_n \cdot v_n \qquad (2)$$

Unfortunately, not all concepts of value are reliable guides to rational decision making. Take moral value, for instance. If a billionaire decides to donate his entire fortune to charity, the expected moral value of doing so might be very high. However, this is because many poor people would benefit from the money, not because the billionaire himself would be any happier. (By assumption, this billionaire is very greedy!) The expected moral value of donating a fortune is far higher than the sort of personal value decision theorists are primarily concerned with. In most cases, moral value is not the sort of value that decision theorists think we should base instrumental, ends-means reasoning on. Therefore, in order to single out the kind of value that is the primary object of study in decision theory – the value of an outcome as evaluated from the decision maker's point of view – it is helpful to introduce the concept of *utility*. Utility is an abstract entity that cannot be directly observed. By definition, the utility of an outcome depends on how valuable the outcome is from the decision maker's point of view. The principle of maximising expected utility is obtained from the principle of maximising expected value by replacing v for u in equation (2).

$$EU = p_1 \cdot u_1 + p_2 \cdot u_2 + \cdots + p_n \cdot u_n \tag{3}$$

In the remainder of this chapter we shall focus on the principle of maximising expected utility, rather than any other versions of the expectation thesis. It is worth noticing that the expected utility principle can be applied also in cases in which outcomes are non-monetary. Consider the following example. David and his partner Rose are about to deliver a sailing yacht from La Coruña (in the north of Spain) to English Harbour, Antigua (in the West Indies). Because of the prevailing weather systems, there are only two feasible routes across the Atlantic, either a direct northern route or a slightly longer southern route. Naturally, the couple wish to cross the Atlantic as quickly as possible. The number of days required for the crossing depends on the route they choose and the meteorological situation. Weather-wise, the decisive factor is whether or not a high pressure zone develops over the Azores after they have set off from the coast of Spain. There are reliable meteorological data going back more than a hundred years, and the probability that a high pressure zone will develop is 83%. By studying the meteorological data and the charts, they figure out that the decision problem they are facing is that shown in Table 4.2.

Since David and Rose wish to make the crossing in as few days as possible, the utility of the outcomes is *negatively correlated* with the number of days at sea. Hence, the utility of sailing for 27 days, which we write as $u(27)$, is lower than the utility of sailing for 18 days, $u(18)$. For simplicity, we assume that in this particular case the utility function is linear with respect

Table 4.2

	High pressure zone over the Azores (83%)	No high pressure zone over the Azores (17%)
Northern route	27 days	14 days
Southern route	18 days	21 days

to the number of days spent at sea. It then follows that the expected utilities of the two alternatives are as follows.

EU(Northern route) $= 0.83 \cdot u(27) + 0.17 \cdot u(14) = u(24.79)$

EU(Southern route) $= 0.83 \cdot u(18) + 0.17 \cdot u(21) = u(18.51)$

Clearly, David and Rose ought to choose the southern route, since $u(18.51) > u(24.79)$, according to our assumption about the correlation between utility and the number of days spent at sea.

Box 4.1   A risky decision

Placebo Pharmaceuticals is eager to expand its product portfolio with a drug against cardio-vascular diseases. Three alternative strategies have been identified. The first is to hire a research team of 200 people to develop the new drug. However, to develop a new drug is expensive (about $50M) and it is also far from certain that the team will manage to successfully develop a drug that meets the regulatory requirements enforced by the Food and Drug Administration; the probability is estimated to be about one in ten. The second alternative is to acquire a small company, Cholesterol Business Inc., that has already developed a drug that is currently undergoing clinical trials. If the trials are successful the Food and Drug Administration will of course license the product rather rapidly, so this alternative is more likely to be successful. According to the executive director of Placebo Pharmaceuticals the probability of success is about 0.8. The downside is that the cost of taking over Cholesterol Inc. is very high, about $120M, since several other big pharmaceutical companies are also eager to acquire the company. The third alternative is to simply buy a licence for $170M from a rival company to produce and market an already existing drug. This is the least risky option, since the board of Placebo Pharmaceuticals knows for sure for what it pays. Finally, to complete the list of alternatives also note that there is a fourth alternative: to do nothing and preserve the status quo.

In order to make a rational decision, Placebo Pharmaceuticals decides to hire a decision analyst. After conducting series of interviews with the board members the decision analyst is able to establish that Placebo Pharmaceuticals' utility of a profit greater than zero is linear and directly proportional to the profit, whereas its disutility of losses L (i.e. a profit equal to or smaller than zero) is determined by the formula $u = 2 \cdot L$. The

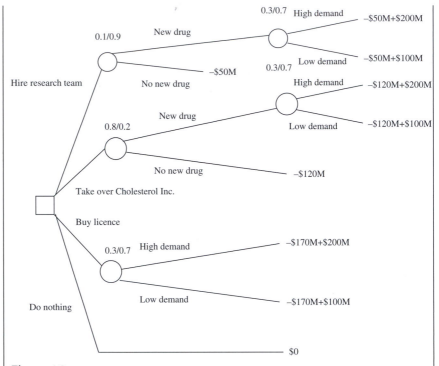

Figure 4.2

decision analyst also concludes that the revenue will depend on future demand of the new drug. The probability that demand will be high is 0.3, in which case the revenue will be $200 M. The probability that demand is low is 0.7, and the revenue will then be $100 M. To explain her findings to the executive director the decision analyst draws the decision tree shown in Figure 4.2.

It can be easily verified that the four alternatives illustrated in the decision tree may lead to nine different outcomes.

Hire research team:	+$150 M with a probability of 0.1 · 0.3
	+$50 M with a probability of 0.1 · 0.7
	−$50 M with a probability of 0.9
Take over Cholesterol Inc.:	+$80 M with a probability of 0.8 · 0.3
	−$20 M with a probability of 0.8 · 0.7
	−$120 M with a probability of 0.2
Buy licence:	+$30 M with a probability of 0.3
	−$70 M with a probability of 0.7
Do nothing:	$0 with a probability of 1

In order to reach a decision, the executive director now applies the principle of maximising expected utility. She recalls that the utility of losses is $u = 2 \cdot L$.

Hire research team:	$+150M \cdot 0.03 + 50M \cdot 0.07 - (2 \cdot 50M) \cdot 0.9$ $= -82M$ utility units
Take over Cholesterol Inc.:	$+80M \cdot 0.24 - (2 \cdot 20M) \cdot 0.56 - (2 \cdot 120M) \cdot 0.2 = -51.2M$ utility units
Buy licence:	$+30M \cdot 0.3 - (2 \cdot 70M) \cdot 0.7 = -89M$ utility units
Do nothing:	0 utility units

Based on the calculations above, the executive director of Placebo Pharmaceuticals decides that the rational thing to do is to do nothing, i.e. to abandon the plan to expand the product portfolio with a drug against cardio-vascular diseases.

4.2 Why is it rational to maximise expected utility?

Decision theorists have proposed two fundamentally different arguments for the expected utility principle. The first argument is based on the law of large numbers; it seeks to show that in *the long run* you will be better off if you maximise expected utility. The second argument aims at deriving the expected utility principle from some more fundamental axioms for rational decision making, which make no reference to what will happen in the long run. We shall return to the axiomatic approach in the next section.

The law of large numbers is a mathematical theorem stating that everyone who maximises expected utility will almost certainly be better off in the long run. In this context the term 'almost certainly' has a very precise meaning. If a random experiment (such as rolling a die or tossing a coin) is repeated n times and each experiment has a probability p of leading to a predetermined outcome, then the probability that the percentage of such outcomes differs from p by more than a very small amount ε converges to 0 as the number of trials n approaches infinity. This holds true for every $\varepsilon > 0$, no matter how small. Hence, by performing the random experiment sufficiently many times, the probability that the average outcome differs from the expected outcome can be rendered *arbitrarily* small.

Imagine, for instance, that you are offered 1 unit of utility for sure or a lottery ticket that will yield either 10 units with a probability of 0.2, or nothing with a probability of 0.8. The expected utility of choosing the lottery ticket is 2, which is more than 1. According to the law of large numbers, you cannot be sure that you will *actually* be better off if you choose the lottery, given that the choice is offered to you only once. However, what you do know for sure is that *if* you face the same decision over and over again, *then* the probability that you will not be better off by choosing the lottery can be made arbitrarily small by repeating the same decision over and over again. Furthermore, if you repeat the decision in question infinitely many times, then the probability that the average utility and the expected utility differ by more than ε units decreases to zero.

Keynes famously objected to the law of large numbers that, "in the long run we are all dead" (1923: 80). This claim is no doubt true, but what should we say about its relevance? Keynes' point was that no real-life decision maker will ever face any decision an infinite number of times; hence, mathematical facts about what would happen after an infinite number of repetitions are therefore of little normative relevance. A different way to express this concern is the following: Why should one care about what *would* happen if some condition were to be fulfilled, given that one knows for sure at the time of making the decision that this condition is certain not to be fulfilled? Personally I think Keynes was right in questioning the relevance of the law of large numbers. But perhaps there is some way in which it could be saved?

It is also worth pointing out that the relevance of the law of large numbers is partially defeated by another mathematical theorem, known as *gambler's ruin*. Imagine that you and I flip a fair coin, and that I pay you $1 every time it lands heads up, and you pay me the same amount when it lands tails up. We both have $1,000 in our pots as we start to play. Now, if we toss the coin *sufficiently* many times each player will at some point encounter a sequence of heads or tails that is longer than he can afford, i.e. longer than the number of one dollar bills in his pot. If you encounter that very long sequence first, you will go bankrupt. Otherwise I will go bankrupt first. It is mathematically impossible that both of us can 'survive' infinitely many rounds of this game, given that each player starts with pots containing finite amounts of money. This means that the law of large numbers guarantees that you will be better off in the long run by maximising expected utility *only if* your initial pot is infinite, which is a rather unrealistic assumption.

An additional worry about the law of large numbers is that it seems perfectly reasonable to question whether decision makers ever face *the same* decision problem several times. Even if you were to play roulette in a Las Vegas casino for weeks, it seems obvious that each time the croupier drops the ball on the roulette wheel she will do it a little bit differently each time, and to some extent it also seems reasonable to claim that the physical constitution of the wheel will change over time, because of dust and wear and tear. Hence, it is not *exactly* the same act you perform every time you play.

That said, for the law of large numbers to work it is strictly speaking not necessary to assume that the agent is facing the same decision problem, in a literal sense, over and over again. All we need to assume is that the *probability* of each outcome is fixed over time. Note, however, that if this is the correct way of understanding the argument, then it will become sensitive to the definition of probability, to be discussed in Chapter 7. According to some theories of probability, the probability that you win when playing roulette is the same over time, but according to other theories this need not necessarily be the case. (For example, if probabilities are defined as relative frequencies, or as subjective degrees of beliefs, then they are very likely to change over time.)

A final worry about the relevance of the law of large numbers is that many decisions under risk are unique in a much stronger sense. It might very well hold true that the probability that John would become happy if he was to marry his partner Joanna is 95%. But so what? He will only marry Joanna once (or at least a very limited number of times). Why pay any attention to the law of large numbers in this decision? The same remark seems relevant in many other unique decisions, i.e. decisions that are made only once, such as a decision to start a war, or appointing a chief executive, or electing a new president. For instance, if the probability is high that the republican candidate will win the next presidential election, the expected utility of investing in the defence industry might be high. However, in this case we cannot justify the expected utility principle by appealing to the law of large numbers, because every presidential election is unique.

4.3 The axiomatic approach

The axiomatic approach to the expected utility principle is not based on the law of large numbers. Instead, this approach seeks to show that the expected utility principle can be derived from axioms that hold independently of what would happen in the long run. If successful, an axiomatic argument can thus overcome the objection that it would not make sense to maximise expected utility in a decision made only once. Here is an extreme example illustrating this point. You are offered to press a green button, and if you do, you will either die or become the happiest person on Earth. If you do not press the button, you will continue to live a rather mediocre life. Let us suppose that the expected utility of pressing the green button exceeds that of not pressing it. Now, what should a rational decision maker do? Axiomatic arguments should entail, if successful, that one should maximise expected utility even in this case, even though the decision is taken only once and the outcome may be disastrous.

Decision theorists have proposed two fundamentally different strategies for axiomatising the expected utility principle. Some axiomatisations are direct, and some are indirect. In the indirect approach, which is the dominant approach, the decision maker does not prefer a risky act to another *because* the expected utility of the former exceeds that of the latter. Instead, the decision maker is asked to state a set of preferences over a set of risky acts. It is irrelevant *how* these preferences are generated. Then, if the set of preferences stated by the decision maker is consistent with a number of structural constraints (axioms), it can be shown that her decisions can be described *as if* she were choosing what to do by assigning numerical probabilities and utilities to outcomes and then maximising expected utility. For an example of such a structural constraint, consider the plausible idea that if act A is judged to be better than act B, then it must not be the case that B is judged to be better than A. Given that this constraint is satisfied, as well as a number of more complex and controversial constraints, it is possible to assign numbers representing hypothetical probabilities and utilities to outcomes in such a way that the agent prefers one act over another if and only if the hypothetical expected utility attributed to that alternative is higher than that of all alternatives. A detailed overview of some influential axiomatic constraints on preferences will be given in Chapters 5 and 7.

In the remainder of this section we focus on the direct approach. It is easier to understand, although it should be stressed that it is less influential in the contemporary literature. The direct approach seeks to generate preferences over acts from probabilities and utilities *directly* assigned to outcomes. In contrast to the indirect approach, it is not assumed that the decision maker has access to a set of preferences over acts before he starts to deliberate. Now, it can be shown that the expected utility principle can be derived from four simple axioms. The presentation given here is informal, but the sceptical reader can rest assured that the argument can be (and has been) formalised.

We use the term utility for referring both to the value of an act and to the value of its outcomes. The aim of the axiomatisation is to show that the utility of an act equals the expected utility of its outcomes. Now, the *first axiom* holds that if all outcomes of an act have utility u, then the utility of the act is u. In Table 4.3 axiom 1 thus entails that the utility of act a_1 is 5, whereas the utility of act a_2 is 7.

The *second axiom* is the dominance principle: If one act is certain to lead to outcomes with higher utilities under all states, then the utility of the former act exceeds that of the latter (and if both acts lead to equal outcomes they have the same utility). Hence, in Table 4.3 the utility of a_2 exceeds that of a_1. Note that this axiom requires that states are causally independent of acts. In Chapter 9 we discuss a type of decision problem for which this assumption does not hold true. The present axiomatisation thus supports the expected utility principle only in a restricted class of decision problems.

The *third axiom* holds that every decision problem can be transformed into a decision problem with equiprobable states, by splitting the original states into parallel ones, without affecting the overall utility of any of the acts in the decision problem; see Table 4.4.

The gist of this axiom is that a_1 and a_2 in the leftmost matrix are exactly as good as a_1 and a_2 in the rightmost matrix, simply because the second

Table 4.3

	s_1	s_2	s_3	s_4	s_5
a_1	5	5	5	5	5
a_2	7	7	7	7	7

Table 4.4

	0.2	0.8		0.2	0.2	0.2	0.2	0.2
a_1	1	3	a_1	1	3	3	3	3
a_2	6	2	a_2	6	2	2	2	2

matrix can be obtained from the first by dividing the set of states corresponding to the outcomes slightly differently.

The *fourth* and last axiom is a trade-off principle. It holds that if two outcomes are equally probable, and if the best outcome is made slightly worse, then this can be compensated for by adding *some* (perhaps very large) amount of utility to the other outcome. Imagine, for instance, that Adam offers you to toss a fair coin. If it lands heads up you will be given 10 units of utility, otherwise you receive 2 units. If you refuse to take part in the gamble you receive 5 units. Before you decide whether to gamble or not, Adam informs you that he is willing to change the rules of the gamble such that instead of giving you 10 units of utility if the coin lands heads up he will give you a little bit less, $10 - \varepsilon_1$, but compensate you for this potential loss by increasing the other prize to $2 + \varepsilon_2$ units (Table 4.5). He adds that you are free to choose the value of ε_2 yourself! The fourth axiom does not say anything about whether you should choose 5 units for sure instead of the gamble yielding either 2 or 10 units of utility, or vice versa. Such choices must be determined by other considerations. The axiom only tells you that there is *some* number $\delta > 0$, such that for all ε_1, $0 \le \varepsilon_1 \le \delta$, there is a number ε_2 such that the trade-off suggested by Adam is unimportant to you, i.e. the utility of the original and the modified acts is the same.

If a sufficiently large value of ε_2 is chosen, even many risk-averse decision makers would accept the suggested trade-off. This means that this axiom can be accepted by more than just decision makers who are neutral to risk-taking. However, this axiom is nevertheless more controversial than the others, because it implies that once ε_1 and ε_2 have been established, these constants can be added over and over again to the utility numbers representing this pair of outcomes. Put in mathematical terms, it is assumed that ε_2 is a function of ε_1, but not of the initial levels of utility. (The axiom can be weakened, however, such that ε_2 becomes a function of more features of the decision problem, but it is beyond the scope of this book to explore this point any further here.)

Table 4.5

	0.5	0.5		0.5	0.5
a_1	5	5	a_1	5	5
a_2	2	10	a_2	$2 + \varepsilon_2$	$10 - \varepsilon_1$

The axioms informally outlined above together entail that the utility of an act equals the expected utility of its outcomes. Or, put in slightly different words, the act that has the highest utility (is most attractive) will also have the highest expected utility, and vice versa. This appears to be a strong reason for letting the expected utility principle guide one's choices in decisions under risk. A more stringent formulation of this claim and a proof is provided in Box 4.2.

Box 4.2   A direct axiomatisation of the expected utility principle

Consider the following four axioms.

EU 1   If all outcomes of an act have utility u, then the utility of the act is u.

EU 2   If one act is certain to lead to better outcomes under all states than another, then the utility of the first act exceeds that of the latter; and if both acts lead to equal outcomes they have the same utility.

EU 3   Every decision problem can be transformed into a decision problem with equally probable states, in which the utility of all acts is preserved.

EU 4   If two outcomes are equally probable, and if the better outcome is made slightly worse, then this can be compensated for by adding some amount of utility to the other outcome, such that the overall utility of the act is preserved.

Theorem 4.1   Let axioms EU 1–4 hold for all decision problems under risk. Then, the utility of an act equals its expected utility.

Proof   The proof of Theorem 4.1 consists of two parts. We first show that $\varepsilon_1 = \varepsilon_2$ (see page 76) whenever EU 4 is applied. Consider the three decision problems in Table 4.6, in which u_1 and u_2 are some utility levels such that u_1 is higher than u_2, while their difference is less than ε_1. (That is, $u_1 - u_2 < \varepsilon_1$.)

Table 4.6

	s	s'		s	s'		s	s'
a_1	u_1	u_2	a_1	u_1	u_2	a_1	u_1	u_2
a_2	u_1	u_2	a_2	$u_1 - \varepsilon_1$	$u_2 + \varepsilon_2$	a_2	$u_1 - \varepsilon_1 + \varepsilon_2$	$u_2 + \varepsilon_2 - \varepsilon_1$

In the leftmost decision problem a_1 has the same utility as a_2, because of EU 2. The decision problem in the middle is obtained by applying EU 4 to act a_2. Note that the utility of both acts remains the same. Finally, the rightmost decision problem is obtained from the one in the middle by applying EU 4 to a_2 again. The reason why ε_1 is subtracted from $u_2 + \varepsilon_2$ is that the utility of the rightmost outcome of a_2 now exceeds that of the leftmost, since the difference between u_1 and u_2 was assumed to be less than ε_1. By assumption, the utility of both acts has to remain the same, which can only be the case if $\varepsilon_1 = \varepsilon_2$. To see why, assume that it is not the case that $\varepsilon_1 = \varepsilon_2$. EU 2 then entails that either a_2 dominates a_1, or a_1 dominates a_2, since $-\varepsilon_1 + \varepsilon_2 = \varepsilon_2 - \varepsilon_1$.

In the second step of the proof we make use of the fact that $\varepsilon_1 = \varepsilon_2$ whenever EU 4 is applied. Let D be an arbitrary decision problem. By applying EU 3 a finite number of times, D can be transformed into a decision problem D* in which all states are equally probable. The utilities of all acts in D* are equal to the utility of the corresponding acts in D. Then, by adding a small amount of utility ε_1 to the lowest utility of a given act and at the same time subtracting the same amount from its highest utility (as we now know we are allowed to do), and repeating this operation a finite number of times, we can ensure that all utilities of each act over all the equally probable states will be equalised. Since all states are equally probable, and we always withdraw and add the same amounts of utilities, the expected utility of each act in the modified decision problem will be exactly equal to that in the original decision problem. Finally, since all outcomes of the acts in the modified decision problem have the same utility, say u, then the utility of the act is u, according to EU 1. It immediately follows that the utility of each act equals its expected utility. $\square$

4.4 Allais' paradox

The expected utility principle is by no means uncontroversial. Naturally, some objections are more sophisticated than others, and the most sophisticated ones are referred to as paradoxes. In the following sections we shall discuss a selection of the most thought-provoking paradoxes. We start with Allais' paradox, which was discovered by the Nobel Prize winning economist Maurice Allais. In the contemporary literature, this paradox is directed both against the expected utility principle in general, as well as against one

Table 4.7

	Ticket no. 1	Ticket no. 2–11	Ticket no. 12–100
Gamble 1	$1M	$1M	$1M
Gamble 2	$0	$5M	$1M
Gamble 3	$1M	$1M	$0
Gamble 4	$0	$5M	$0

of the axioms frequently used in (indirect) axiomatisation of it. In this section we shall conceive of the paradox as a general argument against the expected utility principle. Consider the gambles in Table 4.7, in which exactly one winning ticket will be drawn at random.

In a choice between Gamble 1 and Gamble 2 it seems reasonable to choose Gamble 1 since it gives the decision maker one million dollars for sure ($1M), whereas in a choice between Gamble 3 and Gamble 4 many people would feel that it makes sense to trade a ten-in-hundred chance of getting $5M, against a one-in-hundred risk of getting nothing, and consequently choose Gamble 4. Several empirical studies have confirmed that most people reason in this way. However, no matter what utility one assigns to money, the principle of maximising expected utility recommends that the decision maker prefers Gamble 1 to Gamble 2 if and only if Gamble 3 is preferred to Gamble 4. There is simply no utility function such that the principle of maximising utility is consistent with a preference for Gamble 1 to Gamble 2 *and* a preference for Gamble 4 to Gamble 3. To see why this is so, we calculate the *difference* in expected utility between the two pairs of gambles. Note that the probability that ticket 1 will be drawn is 0.01, and the probability that one of the tickets numbered 2–11 will be drawn is 0.1; hence, the probability that one of the tickets numbered 12–100 will be drawn is 0.89. This gives the following equations:

$$
\begin{aligned}
u(G1) - u(G2) &= u(1M) - [0.01u(0M) + 0.1u(5M) + 0.89u(1M)] \\
&= 0.11u(1M) - [0.01u(0) + 0.1u(5M)]
\end{aligned}
\tag{1}
$$

$$
\begin{aligned}
u(G3) - u(G4) &= [0.11u(1M) + 0.89u(0)] - [0.9u(0M) + 0.1u(5M)] \\
&= 0.11u(1M) - [0.01u(0) + 0.1u(5M)]
\end{aligned}
\tag{2}
$$

Equations (1) and (2) show that the difference in expected utility between G1 and G2 is precisely the same as the difference between G3 and G4. Hence, no matter what the decision maker's utility for money is, it is impossible to simultaneously prefer G1 to G2 *and* to prefer G4 to G3 without violating the expected utility principle. However, since many people who have thought very hard about this example still feel it would be rational to stick to the problematic preference pattern described above, there seems to be something wrong with the expected utility principle.

Unsurprisingly, a number of decision theorists have tried to find ways of coping with the paradox. Savage, a pioneer of modern decision theory, made the following point:

> if one of the tickets numbered from 12 through 100 is drawn, it does not matter, in either situation which gamble I choose. I therefore focus on the possibility that one of the tickets numbered from 1 through 11 will be drawn, in which case [the choice between G1 and G2 and between G3 and G4] are exactly parallel … It seems to me that in reversing my preference between [G3 and G4] I have corrected an error. (Savage 1954: 103)

Savage's point is that it does not matter which alternative is chosen under states that yield the same outcomes, so those states should be ignored. Instead, decision makers should base their decisions entirely on features that differ between alternatives. This idea is often referred to as the sure-thing principle; we will discuss it in more detail in Chapter 7. That said, some people find the sure-thing principle very hard to accept, and argue that this principle is the main target of the paradox. In their view, Savage has failed to explain *why* sure-thing outcomes should be ignored.

Another type of response to Allais' paradox is to question the accuracy of the formalisation of the decision problem. The outcome of getting $0 in G2 is very different from the outcome of getting $0 in G4. The disappointment one would feel if one won nothing instead of a fortune in G2 is likely to be substantial. This is because in the choice between G1 and G2 the first alternative is *certain* to yield a fortune, whereas in the choice between G3 and G4 no alternative is certain to yield a fortune. A more accurate decision matrix would therefore look as in Table 4.8. Note that it no longer holds true that the expected utility principle is inconsistent with the preference pattern people actually entertain.

A drawback of this response is that it seems difficult to tell exactly how fine-grained the description of outcomes ought to be. In principle, it seems that *every* potential violation of the expected utility principle could be

Table 4.8

	Ticket no. 1	Ticket no. 2–11	Ticket no. 12–100
Gamble 1	$1M	$1M	$1M
Gamble 2	$0 and disappointment	$5M	$1M
Gamble 3	$1M	$1M	$0
Gamble 4	$0	$5M	$0

rejected by simply making the individuation of outcomes more fine-grained. However, this would make the principle immune to criticism, unless one has some independent reason for adjusting the individuation of outcomes.

4.5 Ellsberg's paradox

This paradox was discovered by Daniel Ellsberg when he was a Ph.D. student in economics at Harvard in the late 1950s. Suppose the decision maker is presented with an urn containing 90 balls, 30 of which are red. The remaining 60 balls are either black or yellow, but the proportion between black and yellow balls is unknown. The decision maker is then offered a choice between the following gambles:

Gamble 1   Receive $100 if a red ball is drawn
Gamble 2   Receive $100 if a black ball is drawn

When confronted with these gambles you may reason in at least two different ways. First, you may argue that it would be better to choose G1 over G2, since the proportion of red balls is known for sure whereas one knows almost nothing about the number of black balls in the urn. Second, you may believe that there are in fact many more black than red balls in the urn, and therefore choose G2. The paradox will arise no matter how you reason, but for the sake of the argument we assume that you prefer G1 to G2. Now, after having made a choice between G1 and G2 you are presented with a second set of gambles.

Gamble 3   Receive $100 if a red or yellow ball is drawn
Gamble 4   Receive $100 if a black or yellow ball is drawn.

Do you prefer G3 or G4? All four gambles are illustrated in Table 4.9. When confronted with the new pair of gambles, it seems that a person who

Table 4.9

	30	60	
	Red	Black	Yellow
Gamble 1	$100	$0	$0
Gamble 2	$0	$100	$0
Gamble 3	$100	$0	$100
Gamble 4	$0	$100	$100

prefers G1 to G2 is likely to prefer G4 to G3, since G4 is a gamble with *known probabilities*. The probability of winning $100 in G4 is known for sure to be 60/90.

The point of Ellsberg's example is the following. No matter what the decision maker's utility for money is, and no matter what she believes about the proportion of black and yellow balls in the urn, the principle of maximising expected utility can never recommend G1 over G2 *and* G4 over G3, or vice versa. This is because the expected utility of G1 exceeds that of G2 if and only if the expected utility of G3 exceeds that of G4. To show this, we calculate the difference in expected utility between G1 and G2, as well as the difference between G3 and G4. For simplicity, we assume that the utility of $100 equals M and that the utility of $0 equals 0 on your personal utility scale. (Since utility is measured on an interval scale, these assumptions are completely innocent.) Hence, if you believe that there are B black balls in the urn, the difference in expected utility between the gambles is as follows.

$$eu(G1) - eu(G2) = 30/90M - B/90M = 30M - BM$$

$$eu(G3) - eu(G4) = 30/90M + (60 - B)/90M - 60/90M$$
$$= 30M + (60 - B)M - 60M = 30M - BM$$

Note that the paradox cannot be avoided by simply arguing that G2 ought to be preferred over G1, because then it would presumably make sense to also prefer G3 over G4; such preferences indicate that the decision maker seeks to avoid gambles with known probabilities. As shown above, eu(G1) – eu(G2) = eu(G3) – eu(G4), so G2 can be preferred over G1 if and only if G4 is preferred over G3.

The Ellsberg paradox is in many respects similar to the Allais paradox. In both paradoxes, it seems reasonable to violate Savage's sure-thing principle; that is, it does not make sense to ignore entire states just because they have parallel outcomes. However, the *reason* why it seems plausible to take into account outcomes that occur for sure under some states is different. In the Allais paradox, G1 is better than G2 because it guarantees that one gets a million dollars. In the Ellsberg paradox G1 is better than G2 because one knows the exact probability of winning $100 although no alternative is certain to lead to a favourable outcome. Arguably, this shows that the intuitions that get the paradoxes going are fundamentally different. The Ellsberg paradox arises because we wish to avoid epistemic uncertainty about probabilities, whereas the Allais paradox arises because we wish to avoid uncertainty about outcomes.

4.6 The St Petersburg paradox

The St Petersburg paradox is derived from the St Petersburg game, which is played as follows. A fair coin is tossed until it lands heads up. The player then receives a prize worth 2^n units of utility, where n is the number of times the coin was tossed. So if the coin lands heads up in the first toss, the player wins a prize worth 2 units of utility, but if it lands heads up on, say, the fourth toss, the player wins $2^4 = 2 \cdot 2 \cdot 2 \cdot 2 = 16$ units of utility. How much utility should you be willing to 'pay' for the opportunity to play this game? According to the expected utility principle you must be willing to pay any finite amount of utility, because $\frac{1}{2} \cdot 2 + \frac{1}{4} \cdot 4 + \frac{1}{8} \cdot 8 + \cdots = 1 + 1 + 1 + \cdots = \sum_{n=1}^{\infty} \left(\frac{1}{2}\right)^n \cdot 2^n = \infty$. But this is absurd. Arguably, most people would not pay even a hundred units. The most likely outcome is that one wins only a very small amount of utility. For instance, the probability that one wins at most 8 units is $0.5 + 0.25 + 0.125 = 0.875$.

The St Petersburg paradox was discovered by the Swiss mathematician Daniel Bernoulli (1700–1782), who was working in St Petersburg for a couple of years at the beginning of the eighteenth century. The St Petersburg paradox is, of course, not a paradox in a strict logical sense. No formal contradiction is deduced. But the recommendation arrived at, that one should sacrifice any finite amount of utility for the privilege of playing the St Petersburg game, appears to be sufficiently bizarre for motivating the use of the term paradox.

In 1745 Buffon argued in response to Bernoulli that sufficiently improbable outcomes should be regarded as "morally impossible", i.e. beyond concern. Hence, in Buffon's view, a rational decision maker should simply disregard the possibility of winning a very high amount, since such an outcome is highly improbable. Buffon's idea is closely related to the principle of *de minimis* risks, which still plays a prominent role in contemporary risk analysis. (Somewhat roughly put, the *de minimis* principle holds that sufficiently improbable outcomes, such as comet strikes, should be ignored.) From a mathematical point of view, it is obvious that if probabilities below a certain threshold are ignored, then the expected utility of the St Petersburg gamble will be finite. That said, this resolution of course seems to be ad hoc. Why on earth would it be rational to ignore highly improbable outcomes?

Other scholars have tried to resolve the St Petersburg paradox by imposing an upper limit on the decision maker's utility scale. From a historical perspective, this is probably the most prominent resolution of the paradox. The eighteenth-century mathematician Cramer, who discussed Bernoulli's original formulation of the paradox in which the prizes consisted of ducats instead of utility, suggested that "any amount above 10 millions, or (for the sake of simplicity) above $2^{24} = 16677216$ ducats [should] be deemed ... equal in value to 2^{24} ducats" (in Bernoulli 1738/1954: 33). More recently, Nobel Prize winner Kenneth Arrow has maintained that the utility of wealth should be "taken to be a bounded function.... since such an assumption is needed to avoid [the St Petersburg] paradox" (Arrow 1970: 92). In order to see how the introduction of an upper limit affects the paradox, let L be the finite upper limit of utility. Then the expected utility of the gamble would be finite, because:

$$\frac{1}{2} \cdot 2 + \frac{1}{4} \cdot 4 + \frac{1}{8} \cdot 8 + \cdots + \frac{1}{2^k} \cdot L + \frac{1}{2^{k+1}} \cdot L \cdots = \sum_{i=1}^{k-1} (1/2)^i \cdot 2^i + \sum_{i=k}^{\infty} (1/2)^i \cdot L$$

$$= \sum_{i=1}^{k-1} (1/2)^i \cdot 2^i + \left(1 - \sum_{j=1}^{k-1} (1/2)^j \right) \cdot L$$

A common reaction to Arrow's proposal is that the introduction of an upper limit is also ad hoc. Furthermore, even if one could overcome this objection, the introduction of a bounded utility scale may not resolve the paradox anyway. This is because the paradox has little to do with *infinite* utility. Arguably, the paradox arises whenever the expected utility of a gamble is unreasonably high *in comparison to* what we feel would be reasonable to pay for entering the gamble. To see this, note that a slightly modified version of the St Petersburg paradox arises even if only small finite amounts of utility are at stake, as in the following gamble.

> **Gamble 1**   A fair coin is tossed until it lands heads up. The player thereafter receives a prize worth min $\{2^n \cdot 10^{-100}, L\}$ units of utility, where n is the number of times the coin was tossed.

Suppose L equals 1. Now, the expected utility of Gamble 1 has to be greater than $332 \cdot 10^{-100}$ units of utility (since 2^{332} is approximately equal to 10^{100}). However, on average in one out of two times the gamble is played, you win only $2 \cdot 10^{-100}$ units, and in about nine times out of ten you win no more than $8 \cdot 10^{-100}$ units. This indicates that even though the expected utility of Gamble 1 is finite, and indeed very small, it is nevertheless paradoxically high *in comparison to* the amount of utility the player actually wins.

Another resolution of the St Petersburg paradox was suggested by Richard C. Jeffrey. He claimed that, "anyone who offers to let the agent play the St. Petersburg game is a liar, for he is pretending to have an indefinitely large bank" (1983: 154). This is because no casino or bank can possibly fulfil its commitments towards the player in the case that a very large number of tails precedes the first head; hence, the premises of the gamble can never be valid. A possible reply to Jeffrey's argument is to point out that all sorts of prizes should, of course, be allowed. Suppose, for instance, that after having played the St Petersburg gamble you will be connected to Robert Nozick's experience machine. By definition, the experience machine can create any experience in you, e.g. intense happiness or sexual pleasure. The fact that there is a limited amount of money in the world is therefore no longer a problem.

There is also another response to Jeffrey's proposal. As before, the main idea is to show that a slightly modified version of the St Petersburg paradox arises even if we accept Jeffrey's restriction, i.e. if we assume that the amount of utility in the bank is finite. Consider Gamble 1 again, and let L equal the total amount of utility available in the bank. Now, Jeffrey's requirement of a finite amount of utility in the bank is obviously satisfied. Arguably, the most important point is that if a new paradoxical conclusion can be obtained just by making some minor alterations to the original problem, then the old paradox will simply be replaced by a new one, and nothing is gained. The following gamble is yet another illustration of this point, which raises a more general issue that goes beyond Jeffrey's proposal.

> **Gamble 2**   A manipulated coin, which lands heads up with probability 0.4, is tossed until it lands heads up. The player thereafter receives a prize worth 2^n units of utility, where n is the number of times the coin was tossed.

Common sense tells us that Gamble 2 should be preferred to the original St Petersburg gamble, since it is more likely to yield a long sequence of tosses and consequently better prizes. However, the expected utility of both gambles is infinite, because $\sum\limits_{n=1}^{\infty} (0.4)^n \cdot 2^n = \infty$. Hence, the principle of maximising expected utility recommends us to judge both gambles as equally valuable. This is also absurd. Any satisfactory account of rationality must entail that Gamble 2 is better than the original St Petersburg gamble.

4.7 The two-envelope paradox

The two-envelope paradox arises from a choice between two envelopes, each of which contains some money. A trustworthy informant tells you that one of the envelopes contains exactly twice as much as the other, but the informant does not tell you which is which. Since this is all you know you decide to pick an envelope at random. Let us say you pick envelope A. Just before you open envelope A you are offered to swap and take envelope B instead. The following argument indicates that you ought to swap. Let x denote the amount in A. Then envelope B has to contain either $2x$ or $x/2$ dollars. Given what you know, both possibilities are equally likely. Hence, the expected monetary value of swapping to B is $\frac{1}{2} \cdot 2x + \frac{1}{2} \cdot \frac{x}{2} = \frac{5}{4}x$. Since $\frac{5}{4}x$ is more than x, it is rational to take B instead of A.

However, just as you are about to open envelope B, you are offered to swap back. The following argument indicates that you ought to take envelope A. Let y denote the amount in envelope B. It then follows that envelope A contains either $2y$ or $y/2$ dollars. As before, both possibilities are equally likely, so the expected monetary value of swapping is $\frac{1}{2} \cdot 2y + \frac{1}{2} \cdot \frac{y}{2} = \frac{5}{4}y$. Since $\frac{5}{4}y$ is more than y you ought to swap.

Table 4.10

	1/2	1/2	Expected value
Envelope A	x	x	x
Envelope B	$2x$	$1/2x$	$5x/4$

Table 4.11

	1/2	1/2	Expected value
Envelope A	$2y$	$1/2y$	$5y/4$
Envelope B	y	y	y

Clearly, there must be something wrong with the reasoning outlined here. It simply cannot hold true that the expected monetary value of choosing A exceeds that of choosing B, at the same time as the expected monetary value of choosing B exceeds that of choosing A. Consider Table 4.10 and Table 4.11.

The present formulation of the paradox presupposes that there is no upper limit to how much money there is in the world. To see this, suppose that there indeed is some upper limit L to how much money there is in the world. It then follows that no envelope can contain more than $(2/3)L$, in which case the other envelope would be certain to contain $(1/3)L$. (If, say, envelope A contains $(2/3)L$, then it would clearly be false that envelope B contains either $2 \cdot (2/3)L$ or $1/2 \cdot (2/3)L$. Only the latter alternative would be a genuine possibility.) Hence, for the paradox to be viable one has to assume that the amount of money in the world is infinite, which is implausible. That said, the paradox can easily be restated without referring to monetary outcomes; if we assume the existence of infinite utilities the paradox will come alive again.

The two-envelope paradox can also be generated by starting from the St Petersburg paradox: A fair coin is flipped n times until it lands heads up. Then a prize worth 2^n units of utility is put in one of the envelopes and either half or twice that amount in the other envelope. It follows that, for every finite n, if the first envelope contains 2^n units of utility, one always has reason to swap to the other envelope, since its expected utility is higher. However, as we know from the discussion of the St Petersburg paradox, the *expected utility* of the contents in each envelope is infinite.

At present there is no consensus on how to diagnose the two-envelope paradox. A large number of papers have been published in philosophical journals. Most attempt to show that the probability assignments are illegitimate, for one reason or another. However, it has also been argued that the way the outcomes are described does not accurately represent the real decision problem. I leave it to the reader to make up her own mind about this surprisingly deep problem.

Exercises

4.1 Consider the decision problem illustrated below.

	1/2	1/4	1/4
a_1	$49	$25	$25
a_2	$36	$100	$0
a_3	$81	$0	$0

(a) The decision maker's utility u of money is linear. Which act should be chosen according to the principle of maximising expected monetary value?

(b) The decision maker's utility u of money x is given by the formula $u(x) = \sqrt{x}$. Which act should be chosen according to the principle of maximising expected utility?

4.2 I am in my office in Cambridge, but I have to catch a flight from Heathrow this afternoon. I must decide whether to go to Heathrow by coach, which comes relatively cheap at £40, or buy a train ticket for £70. If I take the coach I might get stuck in an intense traffic jam and miss my flight. I would then have to buy a new ticket for £100. According to the latest statistics, the traffic jam on the M25 to Heathrow is intense one day in three. (a) Should I travel by train or coach? (b) This description of my decision problem overlooks a number of features that might be relevant. Which?

4.3 (a) You are in Las Vegas. The probability of winning a jackpot of $350,000 is one in a million. How much should you, who find no reason to reject the principle of maximising expected utility, be prepared to pay to enter this gamble? Your utility of money is $u(x) = \ln(x + 1)$.

(b) This time the probability of winning the jackpot of $350,000 is one in a thousand. How much should you, who find no reason to reject the principle of maximising expected utility, be prepared to pay to enter this gamble? Your utility of money is $u(x) = \ln(x + 1)$.

(c) Why is the difference between the amount you are willing to pay in (a) and (b) so small?

(d) Why did we assume that your utility function is $u(x) = \ln(x + 1)$, rather than just $u(x) = \ln(x)$?

4.4 (a) What is the law of large numbers?

(b) How is the law of large number related to the theorem known as gambler's ruin?

4.5 (a) Explain why Allais' and Ellsberg's paradoxes pose difficulties for the principle of maximising expected utility. (b) Explain the difference between the two paradoxes – they arise for two different reasons.

4.6 Suppose that you prefer Gamble 1 to Gamble 2, and Gamble 4 to Gamble 3. Show that your preferences are incompatible with the principle of maximising expected utility, no matter what your utility of money happens to be.

	1/3	1/3	1/3
Gamble 1	$50	$50	$50
Gamble 2	$100	$50	$0
Gamble 3	$50	$0	$50
Gamble 4	$100	$0	$0

4.7 (a) Explain why the St Petersburg paradox poses a difficulty for the principle of maximising expected utility.

(b) Construct a new version of the St Petersburg paradox, in which the player rolls a six-sided die instead of tossing a coin.

4.8 There is an interesting connection between the St Petersburg and the two-envelope paradox – explain!

Solutions

4.1 (a) a_2 (b) a_1

4.2 (a) If money is all that matters, and my utility of money is linear, I should buy a train ticket. It will cost me £70, but the expected monetary cost of going by coach is £73.33. (b) Arguably, money is not all that matters. For instance, travelling by train is more comfortable and if

I miss the flight I will be severely delayed. You should also attempt to find out the chances of being delayed on the train.

4.3 (a) The expected utility of the gamble is $\dfrac{999999}{10^6} \cdot (\ln 1) + \dfrac{1}{10^6} \cdot (\ln 350001)$

$= 1.27657 \cdot 10^{-5}$. Hence, you should pay at most $e^{1.27657E-05}$

$\approx \$1.000013$.

(b) The expected utility of the gamble is $\dfrac{999}{10^3} \cdot (\ln 1) + \dfrac{1}{10^3} \cdot (\ln 350001)$

$= 0.012765691$. Hence, you should pay at most $e^{0.012765691} \approx \$1.013$.

(c) Because your marginal utility of money is very small money hardly matters to you.

(d) Because ln 0 is undefined.

4.4 See Section 4.2.

4.5 See Sections 4.3 and 4.4.

4.6 We calculate the difference in expected utility between Gamble 1 and Gamble 2, and between Gamble 3 and Gamble 4. Since these are equal, you have to prefer Gamble 1 to Gamble 2 if and only if you prefer Gamble 3 to Gamble 4.

$$
\begin{aligned}
u(G1) - u(G2) &= u(50) - [1/3u(100) + 1/3u(50) + 1/3(0)] \\
&= 2/3u(50) - 1/3u(100) - 1/3(0)
\end{aligned}
\tag{1}
$$

$$
\begin{aligned}
u(G3) - u(G4) &= [2/3u(50) + 1/3u(0)] - [1/3u(100) + 2/3u(0)] \\
&= 2/3u(50) - 1/3u(100) - 1/3(0)
\end{aligned}
\tag{2}
$$

4.7 (a) See Section 4.6.

(b) A fair die is rolled until a six comes up. The player then receives a prize worth 6^n units of utility, where n is the number of times the die was rolled. So if a six comes up the first time the die is rolled, the player wins a prize worth 6 units of utility, but if a six comes up, say, on the fourth toss, the player wins $6^4 = 6 \cdot 6 \cdot 6 \cdot 6 = 1296$ units of utility. How much utility should you be willing to 'pay' for the opportunity to play this game? According to the expected utility principle you must be willing to pay any finite amount of utility, because

$$
\sum_{n=1}^{\infty} \left(\frac{1}{6}\right)^n \cdot 6^n = \frac{1}{6} \cdot 6 + \frac{1}{36} \cdot 36 + \frac{1}{216} \cdot 216 + \cdots = 1 + 1 + 1 + \cdots = \infty.
$$

4.8 See Section 4.7.

5   Utility

It is hard to think of any minimally reasonable decision rule that totally ignores the concept of utility. However, the concept of utility has many different technical meanings, and it is important to keep these different meanings apart. In Chapter 2 we distinguished three fundamentally different kinds of measurement scales. All scales are numerical, i.e. utility is represented by real numbers, but the information conveyed by the numbers depends on which type of scale is being used.

1. Ordinal scales   ('10 is better than 5')
2. Interval scales   ('the difference between 10 and 5 equals that between 5 and 0')
3. Ratio scales   ('10 is twice as valuable as 5')

In ordinal scales, better objects are assigned higher numbers. However, the numbers do not reflect any information about differences or ratios between objects. If we wish to be entitled to say, for instance, that the difference between ten and five units is exactly as great as the difference between five and zero units, then utility has to be measured on an interval scale. Furthermore, to be entitled to say that ten units of utility is twice as much as five, utility must be measured on a ratio scale.

Arguably, utility cannot be directly revealed by introspection. We could of course ask people to estimate their utilities, but answers gathered by this method would most certainly be arbitrary. Some more sophisticated method is needed. So how on earth is it possible to assign precise numbers to outcomes and acts that accurately reflect their value? And how can this process be a scientific one, rather than a process that is merely arbitrary and at best based on educated guesses?

5.1 How to construct an ordinal scale

Let us first show how to construct an ordinal utility scale. To make the discussion concrete, imagine that you have a collection of crime novels, and that you wish to assign ordinal utilities to each book in your collection. To avoid a number of purely mathematical obstacles we assume that the number of books in the collection is finite. Now, it would by no means be unreasonable to ask you to answer the following question: "Which novel do you like the most, *The Hound of the Baskervilles* by Arthur Conan Doyle or *Death on the Nile* by Agatha Christie?" Suppose you answer *Death on the Nile*. This preference can be represented by the symbol $\succ$ in the following way.

(1) *Death on the Nile* $\succ$ *The Hound of the Baskervilles*

Proposition (1) is true if and only if *Death on the Nile* is preferred over *The Hound of the Baskervilles*. This may sound trivial, but what does this mean, more precisely? How do we check if you do in fact prefer *Death on the Nile* to *The Hound of the Baskervilles*? Economists argue that there is an easy answer to this question: Your preferences are *revealed* in your choice behaviour. Therefore, you prefer x to y if and only if you choose x over y whenever given the opportunity. The main advantage of this proposal is that it links preference to directly observable behaviour, which entails that the concept of preference (and hence utility) becomes firmly connected with empirical observations. However, it is of course easy to question this alleged connection between choice and preference. Perhaps you actually preferred x over y, but chose y by mistake, or did not know that y was available. Furthermore, using the behaviourist interpretation of preferences it becomes difficult to distinguish between strict preference ('strictly better than') and indifference ('equally good as'). The observation that you repeatedly choose x over y is equally compatible with the hypothesis that you strictly prefer x over y as with the hypothesis that you are indifferent between the two. However, for the sake of the argument we assume that the decision maker is able, in one way or another, to correctly state pairwise preferences between any pair of objects.

Indifference will be represented by the symbol $\sim$. For future reference we also introduce the symbol $\succeq$, which represents the relation 'at least as preferred as'. We now have three different preference relations, $\succ$, $\sim$ and $\succeq$. Each of them can easily be defined in terms of the others.

(2a) $x \succeq y$ if and only if $x \succ y$ or $x \sim y$

(2b) $x \sim y$ if and only if $x \succeq y$ and $y \succeq x$

(2c) $x \succ y$ if and only if $x \succeq y$ and not $x \sim y$

Let us return to the example with crime novels. The entire collection of novels can be thought of as a set $B = \{x, y, z, ...\}$, where book x is *Death on the Nile* and book y is *The Hound of the Baskervilles*, and so on. Since you were able to state a preference between *Death on the Nile* and *The Hound of the Baskervilles* it seems reasonable to expect that you are able to compare any two books in your collection. That is, for every x and y in B, it should hold that:

Completeness $x \succ y$ or $x \sim y$ or $y \succ x$.

Completeness rules out the possibility that you fail to muster any preference between some pairs of books. This may sound trivial, but it is in fact a rather strong assumption. To see this, consider a completely different kind of choice: Do you prefer to be satisfied but stupid like a pig, or dissatisfied but clever as Socrates? A natural reaction is to say that the two alternatives are incomparable. (J.S. Mill famously had a different opinion; he preferred the latter to the former.) The possibility of incomparability is not consistent with the completeness axiom.

We moreover assume that strict preferences are *asymmetric* and *negatively transitive*.

Asymmetry	If $x \succ y$, then it is false that $y \succ x$.
Transitivity	If $x \succ y$ and $y \succ z$, then $x \succ z$.
Negative transitivity	If it is false that $x \succ y$ and false that $y \succ z$, then it is false that $x \succ z$.

From asymmetry we can conclude that if *Death on the Nile* is preferred over *The Hound of the Baskervilles*, then it is not the case that *The Hound of the Baskervilles* is preferred over *Death on the Nile*. Transitivity is a slightly more complex property: if *Death on the Nile* is preferred over *The Hound of the Baskervilles* and *The Hound of the Baskervilles* is preferred over *Sparkling Cyanide*, then *Death on the Nile* is preferred over *Sparkling Cyanide*. This assumption has been seriously questioned in the literature, for reasons that we will return to in Chapter 8. Here, we shall accept it without further ado.

Negative transitivity is a slightly stronger version of transitivity. The reason why we assume negative transitivity rather than just transitivity is because we need it. Negative transitivity implies that indifference is transitive (if $x \sim y$ and $y \sim z$, then $x \sim z$), but this does not follow from the assumption that strict preference is transitive. (I leave it to the reader to verify this claim.) Furthermore, it can also be shown that negative transitivity, but not transitivity, entails that $x \succ z$ if and only if, for all y in B, $x \succ y$ or $y \succ z$.

Let us now return to the problem of constructing an ordinal utility scale. What assumptions must we make about preference relations for this to be possible? That is, what must we assume about preferences for there to exist a function u that assigns real numbers to all books in your collection such that better books are assigned higher numbers, i.e.

(3) $x \succ y$ if and only if $u(x) > u(y)$?

Note that there would not, for instance, exist a utility function u if the strict preference relation is cyclic, i.e. if $x \succ y$ and $y \succ z$ and $z \succ x$. There are simply no real numbers that can represent this ordering such that better objects are represented by higher numbers. However, it can be shown that such numbers do exist, i.e. that there is a real-valued function u such that (3) holds true, if and only if the relation $\succ$ is complete, asymmetric and negatively transitive. (A proof is given in Box 5.1.) This means that we have solved the problem we set out to solve – we now know under what conditions it is possible to construct an ordinal utility scale, and how to do it!

5.2 von Neumann and Morgenstern's interval scale

In many cases ordinal utility scales do not provide the information required for analysing a decision problem. The expected utility principle as well as several other decision rules presuppose that utility is measured on an interval scale. In the second edition of the *Theory of Games and Economic Behavior*, published in 1947, John von Neumann and Oskar Morgenstern proposed a theory that has become the default strategy for constructing interval scales, with which other theories of utility must compete. This is partly because von Neumann and Morgenstern's theory does not merely explain what utility is and how it can be measured; their theory also offers an indirect but elegant justification of the principle of maximising expected utility.

The key idea in von Neumann and Morgenstern's theory is to ask the decision maker to state a set of preferences over risky acts. These acts are called lotteries, because the outcome of each act is assumed to be randomly determined by events (with known probabilities) that cannot be

Box 5.1   An ordinal utility scale

Theorem 5.1   Let B be a finite set of objects. Then, $\succ$ is complete, asymmetric and negatively transitive in B if and only if there exists a real-valued function u such that condition (3) on page 94 holds.

Proof   We have to prove both directions of the biconditional. Let us first show that *if* there exists a function u such that (3) holds, *then* the preference relation $\succ$ is complete, asymmetric and negatively transitive. Completeness follows directly from the corresponding property of the real numbers: Since it is always true that $u(x) > u(y)$ or $u(x) = u(y)$ or $u(y) > u(x)$, it follows that $x \succ y$ or $x \sim y$ or $y \succ x$. Furthermore, note that if $u(x) > u(y)$, then it is not the case that $u(y) > u(x)$; this is sufficient for seeing that $\succ$ is asymmetric. Finally, negative transitivity follows from the observation that if $u(x) \geq u(y)$ and $u(y) \geq u(z)$, then $u(x) \geq u(z)$. (Since $x \succ y$ is not compatible with $y \geq x$, given completeness.)

Next, we wish to show that *if* the preference relation $\succ$ is complete, asymmetric and transitive, then there exists a function u such that (3) holds. We do this by outlining one of the many possible ways in which u can be constructed. Before we start, we label the elements in B with index numbers. This can be done by just adding indices to the elements of B, i.e. $B = \{x_1, y_2, z_3 \ldots\}$. The next step is to define the set of elements that are *worse than* an arbitrarily chosen element x. Let $W(x) = \{y : x \succ y\}$, i.e. y is an element in W that is worse than x. Furthermore, let $N(x)$ be the set of index numbers corresponding to W, i.e. $N(x) = \{n : x_n \in W(x_n)\}$.

The utility function u can now be defined as follows: $u(x) = \sum\limits_{n \in N(x)} \left(\frac{1}{2}\right)^n$.

All that remains to do is to verify that u satisfies (3), i.e. that $x \succ y$ if and only if $u(x) > u(y)$. From left to right: Suppose that $x \succ y$. Then, since $\succ$ is transitive and asymmetric, $W(x)$ must contain at least one element that is not in $W(y)$. Hence, $N(x)$ must also contain at least one element that is not in $N(y)$. By stipulation we now have:

$$u(x) = \sum_{n \in N(x)} \left(\frac{1}{2}\right)^n = \sum_{n \in N(y)} \left(\frac{1}{2}\right)^n + \sum_{n \in N(x) - N(y)} \left(\frac{1}{2}\right)^n > \sum_{n \in N(y)} \left(\frac{1}{2}\right)^n = u(y)$$

From right to left: Suppose that $u(x) > u(y)$. Completeness guarantees that $x \succ y$ or $x \sim y$ or $y \succ x$. The last alternative is impossible, since it has already been shown above that if $y \succ x$ then $u(y) > u(x)$, which is a contradiction. For the same reason, we can rule out $x \sim y$, because $u(x) = u(y)$ is also inconsistent with $u(x) > u(y)$. Hence, we can conclude that $x \succ y$. □

controlled by the decision maker. The set of preferences over lotteries is then used for calculating utilities by reasoning 'backwards'. As an illustration of this, we shall consider a simple example. Note that the example only gives a partial and preliminary illustration of von Neumann and Morgenstern's theory.

Suppose that Mr Simpson has decided to go to a rock concert, and that there are three bands playing in Springfield tonight. Mr Simpson thinks that band A is better than B, which is better than C. For some reason, never mind why, it is not possible to get a ticket that entitles him to watch band A or C with 100% certainty. However, he is offered a ticket that entitles him to a 70% chance of watching band A and a 30% chance of watching band C. The only other ticket available entitles him to watch band B with 100% certainty. For simplicity, we shall refer to both options as lotteries, even though only the first involves a genuine element of chance. After considering his two options carefully, Mr Simpson declares them to be equally attractive. That is, to watch band B with 100% certainty is, for Mr Simpson, exactly as valuable as a 70% chance of watching band A and a 30% chance of watching band C. Now suppose we happen to know that Mr Simpson always acts in accordance with the principle of maximising expected utility. By reasoning backwards, we can then figure out what his utility for rock bands is, i.e. we can determine the values of $u(A)$, $u(B)$ and $u(C)$. Consider the following equation, which formalises the hypothesis that Mr Simpson accepts the expected utility principle and thinks that the two options are equally desirable:

$$0.7 \cdot u(A) + 0.3 \cdot u(C) = 1.0 \cdot u(B) \tag{1}$$

Equation (1) has three unknown variables, so it has infinitely many solutions. However, if utility is measured on an interval scale, Equation (1) nevertheless provides all information we need, since the unit and end points of an interval scale can be chosen arbitrarily. (See Chapter 2.) We may therefore stipulate that the utility of the best outcome is 100 (that is, $u(A) = 100$), and that the utility of the worst outcome is 0 (that is, $u(C) = 0$). By inserting these arbitrarily chosen end points into Equation (1), we get the following equation.

$$0.7 \cdot 100 + 0.3 \cdot 0 = 1.0 \cdot u(B) \tag{2}$$

Equation (2) has only one unknown variable, and it can be easily solved: As you can see, $u(B) = 70$. Hence, we now know the following.

$u(A) = 100$
$u(B) = 70$
$u(C) = 0$

Now suppose that Mr Simpson is told that another rock band will play in Springfield tonight, namely band D, which he thinks is slightly better than B. What would Mr Simpson's numerical utility of watching band D be? For some reason, never mind why, Mr Simpson is offered a ticket that entitles him to watch band D with probability p and band C with probability $1 - p$, where p is a variable that he is free to fix himself. To figure out what his utility for D is, he asks himself the following question: "Which value of p would make me feel totally indifferent between watching band B with 100% certainty, and D with probability p and C with probability $1 - p$?" After considering his preferences carefully, Mr Simpson finds that he is indifferent between a 100% chance of watching band B and a 78% chance of watching D combined with a 22% chance of watching C. Since we know that $u(B) = 70$ and $u(C) = 0$, it follows that:

$$1.0 \cdot 70 = 0.78 \cdot u(D) + 0.22 \cdot 0 \tag{3}$$

By solving this equation, we find that $u(D) = 70/0.78 \approx 89.7$. Of course, the same method could be applied for determining the utility of every type of good. For instance, if Mr Simpson is indifferent between a 100% chance of watching rock band B and a 95% chance of winning a holiday in Malibu combined with a 5% chance of watching band C, then his utility of a holiday in Malibu is $70/0.95 = 73.7$.

An obvious problem with this *preliminary* version of von Neumann and Morgenstern's theory is that it presupposes that the decision maker chooses in accordance with the principle of maximising expected utility. We seem to have no reason for thinking that the decision maker will apply the expected utility principle, rather than some other principle, for evaluating lotteries. This very strong assumption needs to be justified in one way or another. Von Neumann and Morgenstern proposed a clever way of doing that. Instead of directly assuming that the decision maker will always apply the expected utility principle (as we did above) they proposed a set of *constraints on rational preferences* which imply that the decision maker behaves *as if* he or she is making decisions by calculating expected utilities. More precisely put, von Neumann and Morgenstern were able to prove that if a decision maker's preferences over the sort of lotteries exemplified above satisfy a number of formal constraints, or *axioms*, then the decision maker's choices can be represented by a function that assigns utilities to lotteries (including lotteries comprising no uncertainty), such that one lottery is preferred to another in the case that the expected utility of the first lottery exceeds that of the latter. In the remaining paragraphs of this section we shall spell out the technical assumptions underlying von Neumann and Morgenstern's theory in more detail, and thereafter prove a version of their so-called representation and uniqueness theorem in Appendix B.

We assume that Z is a finite set of basic prizes, which may include a holiday in Malibu, a ticket to a rock concert, as well as almost any kind of good. That is, the elements of Z are the kind of things that typically constitute outcomes of risky decisions. We furthermore assume that L is the set of lotteries that can be constructed from Z by applying the following inductive definition. (Note that even a 100% chance of winning a basic prize counts as a 'lottery' in this theory.)

1. Every basic prize in Z is a lottery.
2. If A and B are lotteries, then so is the prospect of getting A with probability p and B with probability $1 - p$, for every $0 \leq p \leq 1$.
3. Nothing else is a lottery.

For simplicity, the formula ApB will be used as an abbreviation for a lottery in which one wins A with probability p and B with probability $1 - p$. Thus, the second condition stated above could equally well be formulated as follows: If A and B are lotteries, then so is ApB, for every $0 \leq p \leq 1$. Furthermore, since ApB is a lottery it follows that also $Cq(ApB)$, $0 \leq q \leq 1$ is a lottery, given that q is a probability and C is a lottery. And so on and so forth.

The next assumption introduced by von Neumann and Morgenstern holds that the decision maker is able to state pairwise preferences between lotteries. The formula $A \succ B$ means that lottery A is preferred over lottery B, and $A \sim B$ means that lottery A and B are equally preferred. Now, it should be obvious that preferences have to satisfy *some* structural conditions. For example, it would make little sense to prefer A to B *and* B to A; that is, for a rational decision maker it cannot hold true that $A \succ B$ and $B \succ A$. This property of rational preferences is usually referred to as the asymmetry condition. What further conditions could we impose upon rational preferences? von Neumann and Morgenstern, as well as many other decision theorists, take completeness and transitivity to be two intuitively reasonable conditions. We recognise them from the discussion of ordinal utility. However, note that the objects over which one is suppose to state preferences is now a set of lotteries, not a set of certain outcomes.

vNM 1 (Completeness) $A \succ B$ or $A \sim B$ or $B \succ A$
vNM 2 (Transitivity)   If $A \succ B$ and $B \succ C$, then $A \succ C$

To state the next axiom, let p be some probability strictly greater than zero.

vNM 3 (Independence) $A \succ B$ if and only if $ApC \succ BpC$

The independence axiom is best illustrated by considering an example. Imagine that you are offered a choice between lotteries A and B in Table 5.1. Each ticket is equally likely to be drawn, so the probability of winning, say, \$5M if lottery B is chosen is 10/11.

Suppose that you prefer lottery A to lottery B. Then, according to the independence axiom it must also hold that you prefer the first lottery to the second in the situation illustrated in Table 5.2. That is, you must prefer ApC to BpC.

Table 5.1

	Ticket no. 1	Ticket no. 2–11
A	\$1M	\$1M
B	\$0	\$5M

Table 5.2

	Ticket no. 1	Ticket no. 2–11	Ticket no. 12–100
ApC	\$1M	\$1M	\$1M
BpC	\$0	\$5M	\$1M

Table 5.3

	Ticket no. 1	Ticket no. 2–11	Ticket no. 12–100
ApC	\$1M	\$1M	\$0
BpC	\$0	\$5M	\$0

The independence axiom can be criticised by using the Allais paradox, discussed in Chapter 4. To see this, note that the independence axiom is supposed to hold no matter what C is. Thus, if you prefer A to B in the first situation you must also prefer ApC to BpC in the situation illustrated in Table 5.3. However, in this case it seems entirely reasonable to hold the opposite preference, that is, to prefer BpC to ApC. This is because there is no longer any alternative that gives you a million for sure; hence, it might be worth taking a slightly larger risk and hope to get \$5M instead.

However, despite the worries raised above we shall nevertheless suppose that the independence axiom can be defended in some way or another. This is because we need it for formulating the utility theory proposed by von Neumann and Morgenstern.

The fourth and last axiom proposed by von Neumann and Morgenstern is a continuity condition. Let p and q be some probabilities strictly greater than 0 and strictly smaller than 1.

vNM 4 (Continuity) *If $A \succ B \succ C$ then there exist some p and q such that*
$$ApC \succ B \succ AqC$$

The following example explains the assumption articulated by the continuity axiom. Suppose that A is a prize worth \$10M, B a prize worth \$9M and C a prize worth \$0. Now, according to the continuity axiom, it holds that if you prefer \$10M to \$9M and \$9M to \$0, then there must be some probability p, which may be very close to 1, such that you prefer \$10M with probability p and \$0 with probability $1-p$ over \$9M for certain. Furthermore, there must be some probability q such that you prefer \$9M for certain over \$10M with probability q and \$0M with probability $1 - q$. Of course, some people might feel tempted to deny that these probabilities exist; perhaps it could be argued that there is no probability p simply because \$9M for certain is always better than a lottery yielding either \$10M or \$0, no matter how small the probability of getting \$0 is. The standard reply to this complaint is that p might lie very close to 1.

In addition to the four axioms stated above, we also need to make an additional technical assumption, saying that the probability calculus applies to lotteries. (In some presentations this condition is listed as a separate axiom.) The essence of this assumption is that it does not matter if you are awarded prize A if you first roll a die and then roll it again, or make a double roll, provided that you only get the prize if you get two sixes. Put into mathematical vocabulary, compound lotteries can always be reduced to simple ones, involving only basic prizes. Hence, if p, q, r and s are probabilities such that $pq + (1 - p)r = s$, then $(AqB)p(ArB) \sim AsB$.

The axioms stated above imply the following theorem, which is frequently referred to as von Neumann and Morgenstern's theorem. It consists of two parts, a *representation* part and a *uniqueness* part.

Theorem 5.2   The preference relation $\succ$ satisfies vNM 1–4 if and only if there exists a function u that takes a lottery as its argument and returns a real number between 0 and 1, which has the following properties:

(1) $A \succ B$ if and only if $u(A) > u(B)$.
(2) $u(ApB) = pu(A) + (1 - p)u(B)$.
(3) For every other function u' satisfying (1) and (2), there are numbers $c > 0$ and d such that $u' = c \cdot u + d$.

Property (1) articulates the fact that the utility function u assigns higher utility numbers to better lotteries. From property (1) it follows that $A \sim B$ if and only if $u(A) = u(B)$. Here is a proof: To prove the implication from left to right, suppose for reductio that $A \sim B$ and that $u(A) > u(B)$. It then follows from property (1) that $A \succ B$, which contradicts the completeness axiom. Moreover, if $A \sim B$ and $u(B) > u(A)$ it follows that $B \succ A$, which also contradicts the completeness axiom. Hence, if $A \sim B$ then it has to be the case that $u(A) = u(B)$. Furthermore, to prove the implication from right to left, suppose that $u(A) = u(B)$ and that it is false that $A \sim B$. The completeness axiom then entails that either $A \succ B$ or $B \succ A$ (but not both), and in conjunction with property (1) both possibilities give rise to a contradiction, because neither $u(A) > u(B)$ nor $u(B) > u(A)$ is consistent with the assumption that $u(A) = u(B)$.

Property (2) of the theorem is the expected utility property. It shows us that the value of a compound lottery equals the expected utility of its components. This means that anyone who obeys the four axioms acts *in accordance with* the principle of maximising expected utility. Of course, it does not follow that the decision maker *consciously applies* the expected utility principle. All that follows is that it is possible to rationalise the decision maker's behaviour by pointing out that he acts *as if* he or she were ascribing utilities to outcomes and calculating the expected utilities.

Properties (1) and (2) are the representation part of the theorem, which show how a utility function can be used for *representing* the decision maker's behaviour. Property (3) is the uniqueness part, telling us that all utility functions satisfying (1) and (2) have something important in common: they are all positive linear transformations of each other. This means that every utility function satisfying (1) and (2) can be obtained from every other such function by multiplying the latter by a constant and adding another constant. As explained in Chapter 2, this means that in von Neumann and Morgenstern's theory utility is measured on an interval scale. A proof of von Neumann and Morgenstern's theorem is given in Appendix B.

Commentators have outlined at least three general objections to the significance of von Neumann and Morgenstern's result. If correct, these objections show that a utility function cannot be constructed in the way proposed by them.

(1) *The axioms are too strong.* As pointed out above, the axioms on which the theory relies are not self-evidently true. It can be questioned whether rational agents really have to obey these axioms. We shall return to this criticism, which is by far the most common one, in Chapter 8.

(2) *No action guidance.* By definition, a rational decision maker who is about to choose among a large number of very complex acts (lotteries) has to know already from the beginning which risky act (lottery) to prefer. This follows from the completeness axiom. Hence, a utility function derived in von Neumann and Morgenstern's theory cannot be used by the decision maker for first *calculating* expected utilities and thereafter *choosing* an act having the highest expected utility. The output of von Neumann and Morgenstern's theory is not a set of preferences over acts – on the contrary, these preferences are used as input to the theory. Instead, the output is a (set of) utility function(s) that can

be used for describing the agent as an expected utility maximiser. Hence, ideal agents do not prefer an act *because* its expected utility is favourable, but can only be described *as if* they were acting from this principle. To some extent, the theory thus puts the cart before the horse.

In reply to this objection, it could be objected that someone who is not fully rational (and thus does not have a complete preference ordering over lotteries) might nevertheless get some help from the axioms. First, they can be used for detecting any inconsistencies in the decision maker's set of preferences. Second, once your utility function has been established the expected utility principle could be used for filling any 'missing gaps', i.e. lotteries you have not yet formed preferences about. Note that both these responses presuppose that the decision maker is a non-ideal agent. But what about ideal decision makers? Does it really make sense to *define* an ideal decision maker such that it becomes trivially true that ideal decision makers do not need any action guidance?

(3) *Utility without chance.* It seems rather odd from a linguistic point of view to say that the *meaning* of utility has something to do with preferences over lotteries. For even a decision maker who (falsely) believes that he lives in a world in which every act is certain to result in a known outcome, i.e. a world that is fully deterministic and known to be so, can meaningfully say that the utility of some events exceed that of others. In everyday contexts the concept of utility has no conceptual link to the concept of risk. Hence, it might be questionable to develop a technical notion of utility that presupposes such a link, at least if it is meant to be applied in normative contexts. (Perhaps it might be fruitful for descriptive purposes.)

The obvious reply to this objection is that von Neumann and Morgenstern's theory is not a claim about the meaning of utility, it is a claim about how to measure utility. However, this is not a very helpful reply, because it then follows that their theory is at best a partial theory. If we acknowledge that it would make sense to talk about utilities even if the world was fully deterministic and known to be so, it seems to follow that we would then have to come up with some other method for measuring utility in that world. And if such a method exists, why not use it everywhere?

Box 5.2   How to buy a car from a friend without bargaining

Joanna has decided to buy a used car. Her best friend Sue has a yellow Saab Aero Convertible that she is willing to sell to Joanna. However, since Joanna and Sue are good friends and do not want to jeopardise their friendship, they feel it would be unethical to bargain about the price. Instead, they agree that an ethically fair price of the Saab would be the price at which Joanna is *indifferent* between the Saab and the amount of money paid for the car, irrespective of Sue's preference. Hence, if Joanna is indifferent between the Saab and $10,000 it follows that $10,000 is a fair price of the car. Unfortunately, since Joanna and Sue are friends, Joanna is not able to honestly and sincerely make direct comparisons between the Saab and various amounts of money. When asked to state a preference between the car and some amount of money she cannot tell what she prefers. Furthermore, it does not help to observe her choice behaviour; in a case like this Joanna's spontaneous choices can very well be triggered by random process, rather than robust choice dispositions. Therefore, to overcome Joanna's inability to directly compare the Saab with money they decide to proceed as follows:

1. Sue offers Joanna to state preferences over a large number of hypothetical car lotteries. The prizes include a Ford, the Saab and a Jaguar. It turns out that Joanna's preferences over car lotteries satisfy the von Neumann–Morgenstern axioms and that she is indifferent between getting the Saab for certain and a lottery in which the probability is 0.8 that she wins a Ford and 0.2 that she wins a Jaguar. The von Neumann–Morgenstern theorem (Theorem 5.2) now implies that:

   (i) $0.8 \cdot u(\text{Ford}) + 0.2 \cdot u(\text{Jaguar}) = 1.0 \cdot u(\text{Saab})$

   By letting $u(\text{Jaguar}) = 1$ and $u(\text{Ford}) = 0$ it follows that $u(\text{Saab}) = 0.8$

2. Next, Sue helps Joanna to establish a separate utility function of money by offering her a second set of hypothetical lotteries. This utility function has no (direct) relation to her utility function of cars. As before, Joanna's preferences satisfy the von Neumann–Morgenstern axioms, and for future reference we note that she is indifferent between getting (i) $25,000 for certain and (ii) a lottery in which the probability is 0.6 that she wins $60,000 and 0.4 that she wins nothing. Joanna is also indifferent between getting (i) $60,000 for certain and

(ii) a lottery in which the probability is 0.2 that she wins $25,000 and 0.8 that she wins $90,000. Hence, we have:

(ii) $0.6 \cdot u(\$60,000) + 0.4 \cdot u(\$0) = 1.0 \cdot u(\$25,000)$

(iii) $0.2 \cdot u(\$25,000) + 0.8 \cdot u(\$90,000) = 1.0 \cdot u(\$60,000)$

Equation (ii) implies that if $u(\$60,000) = 1$ and $u(\$0) = 0$, then $u(\$25,000) = 0.6$. Furthermore, according to equation (iii), if $u(\$90,000) = 1$ and $u(\$25,000) = 0$, then $u(\$60,000) = 0.8$. The two utility scales derived from (ii) and (iii) are not directly connected to each other. However, they can of course be merged into a single scale by observing that the difference in utility between $60,000 and $25,000 corresponds to 0.4 units on scale (ii) and 0.8 units on scale (iii). This means that a difference of 0.2 units on scale (iii) between $60,000 and $90,000 would correspond to a difference of 0.1 unit on scale (ii). Hence, $u(\$90,000) = 1.1$ on that scale. The leftmost columns of Table 5.4 summarise the two original scales. The scales to the right show how the original scales can be merged into a single scale as described above, and thereafter calibrated such that the end points become 1 and 0.

3. In the third and final step Joanna has to find a way of connecting her utility scale for cars with her utility scale for money. As mentioned above, she finds it impossible to *directly* compare the Saab with money, but she is willing to compare other cars with money. It turns out that she is indifferent between a Jaguar and $90,000 and between a Ford and $25,000. So on the calibrated scale a Jaguar corresponds to 1 unit of utility and a Ford to 0.55 units. Hence, the difference between a Jaguar and a Ford is $1 - 0.55 = 0.45$ on the calibrated scale. Now recall the first step, in which we established that the utility of the Saab is 80% of the difference between the Jaguar and the Ford. As 80% of 0.45 is 0.36, the utility of the Saab is $0.45 + 0.36 = 0.81$.

Table 5.4

	Scale (ii)	Scale (iii)	Merged scale	Calibrated scale
$90,000		1	1.1	1.1/1.1 = 1
$60,000	1	0.8	1	1/1.1 = 0.91
$25,000	0.6	0	0.6	0.6./1.1 = 0.55
$0	0		0	0/1.1 = 0

What amount of money corresponds to 0.81 units of utility? This is equivalent to asking: What amount $X for certain is judged by Joanna to be exactly as attractive as a lottery in which she wins $90,000 with a probability of 0.81 and $0 with a probability of 0.19? As pointed out in Step 2, Joanna is indeed willing to answer this type of question. After some reflection, she concludes that the amount in question is $55,000. This answers the question we set out to answer at the beginning: An ethically fair price of Sue's yellow Saab Aero Convertible is $55,000.

In summary, this example shows how the von Neumann–Morgenstern theory can be used for making *indirect* comparisons of items the agent is not immediately willing to compare, e.g. a specific car and some amount of money. The trick is to construct two separate scales, one for cars and one for money, and then weld them together into a single scale. So in this type of case Joanna's preferences satisfy the completeness axiom in an *indirect* sense. In Chapter 8, we shall discuss how to reason when the completeness axioms cannot be saved no matter what tricks we employ.

5.3 Can utility be measured on a ratio scale?

Many decision theorists believe that utility can be measured only on an interval scale, and that the best way to construct an interval scale is along the lines suggested by von Neumann and Morgenstern. Arguably, this is a metaphysical claim about the nature of utility. If correct, it tells us something important about what utility is. However, some economists and mathematicians maintain that it is also possible to measure utility on a ratio scale. To render this claim plausible, a radically different approach has to be taken. The point of departure is the observation that there seems to be an intimate link between the utility of an option and the probability with which it is chosen. At first glance this may seem trivial. If you think that $20 is better than $10, then you will of course choose $20 with probability 1 and $10 with probability 0, whenever offered a choice between the two objects. However, the *probabilistic theory* of utility, as I shall call it, is much more sophisticated than that.

To start with, let us suppose that an external observer wishes to figure out what a decision maker's utility function for some set of objects is. We assume that it is not entirely sure, from the observer's perspective, what the decision maker will choose. All the observer can tell is that the probability that some option will be chosen over another is, say, 0.8. Imagine, for instance, that the manager of a supermarket observes that one of her customers tends to buy apples but no bananas eight times out of ten, and bananas but no apples two times out of ten. In that case the manager may conclude that the customer will choose an apple over a banana with a probability of 0.8. Now assume that the observer is somehow able to verify that none of the customer's tastes or desires has changed; that is, the reason why the customer sometimes buys apples and sometimes buys bananas is not that he frequently changes his taste.

Empirical studies show that probabilistic choice behaviour is by no means uncommon, but how can such behaviour be rational? It has been suggested that we may think of the decision maker's choice process as a two-step process. In the first step, the decision maker assesses the utility of each option. Then, in the second step, she makes a choice among the available options by simply trying to choose an alternative that maximises utility. Now, there are at least two possibilities to consider, both of which give rise to probabilistic choice behaviour. First, the decision maker may sometimes fail to choose an option that maximises utility in the second step of the choice process, because she fails to choose the alternative that is best for her. This version of the probabilistic theory is called the *constant utility model*. Second, we may imagine that the decision maker in the second step always chooses an alternative with the highest utility, although the first step is probabilistic, i.e. the probabilities come from the assessment of utilities rather than the choice itself. This version of the probabilistic theory is called the *random utility model*. According to this view one and the same object can be assigned different utilities at two different times, even if everything else in the world is kept constant, because the act of assigning utilities to objects is essentially stochastic.

Now, in order to establish a more precise link between probability and utility, it is helpful to formalise the probabilistic approach. Let x, y, z be arbitrary objects, and let $A, B, \ldots$ be sets of objects, and let the formula $p(x \succ B)$ denote a probability of p that x will be chosen out of B. Hence, $p(apple \succ \{apple, banana\}) = 0.8$ means that the probability is 0.8 that an apple will be chosen over a banana. Furthermore, let $p(A \succ B)$ denote the probability of the following conditional: If A is a subset of B, then the probability is p that the chosen alternative in B is also an element of A.

We now come to the technical core of the probabilistic theory. This is the choice axiom proposed by Duncan Luce. This axiom holds that if A is a subset of B, then the probability that x will be chosen from B equals the probability that x will be chosen from A multiplied by the probability that the chosen alternative in B is also an element of A. In symbols,

Choice axiom   If $A \subset B$, then $p(x \succ B) = p(x \succ A) \cdot p(A \succ B)$.

To grasp what kind of assumption is at stake here it is helpful to consider an example. Suppose that Mr Simpson is visiting a posh restaurant and that he is about to choose a wine from a list containing two red and two white wines. Now, the choice axiom entails that it should not matter if Mr Simpson divides his choice into two stages, that is, first chooses between red and white wine and then between the wines in the chosen subset, or choose directly which of the four wines to order. Hence, if Mr Simpson is indifferent between red and white wine in general, as well as between the two red wines and the two white ones at hand, the probability that a particular bottle will be chosen is $\frac{1}{2} \cdot \frac{1}{2} = \frac{1}{4}$.

Gerard Debreu, winner of the Nobel Prize in economics, has constructed an interesting counter example to the choice axiom: When having dinner in Las Vegas, Mr Simpson is indifferent between seafood and meat, as well as between steak and roast beef. The menu comprises only three dishes: $x =$ lobster, $y =$ steak and $z =$ roast beef. Let B be the entire menu, let A be the set comprising x and y, and let A' be the set of y and z. Then, since Mr Simpson is indifferent between seafood and meat, $p(x \succ B) = \frac{1}{2}$. However, $p(A \succ B) = 1 - p(z \succ B) = 1 - p(z \succ A') \cdot p(A' \succ B) = 1 - \frac{1}{2} \cdot \frac{1}{2} = \frac{3}{4}$. Hence, the choice axiom implies that $p(x \succ B) = p(x \succ A) \cdot p(A \succ B) = \frac{1}{2} \cdot \frac{3}{4} = \frac{3}{8}$.

In response to Debreu's example, people wishing to defend the choice axiom may say that the root of the problem lies in the individuation of alternatives. It could be argued that lobster, steak and roast beef are not alternatives at the same level. Lobster belongs to the category 'seafood', and could equally well be replaced with tuna, or any other seafood dish. However, for the example to work, neither steak nor roast beef can be replaced with some other meat dish, say kebab, because then it would no longer be certain that the agent will remain indifferent between the two meat dishes. Hence, the moral of Debreu's example seems to be that alternatives must be individuated with care. Perhaps the choice axiom should be taken into account already when alternatives are being individuated – it could be conceived of as a normative requirement for how alternatives *ought* to be individuated. We are now in a position to prove the following theorem.

Theorem 5.3   Let B be a finite set of objects such that $p(x \succ B) \neq 0, 1$ for all x in B. Then, if the choice axiom holds for B and all its subsets, and the axioms of probability theory hold, then there exists a positive real-valued function u on B such that for every $A \subset B$ it holds that

(1) $p(x \succ A) = \dfrac{u(x)}{\sum\limits_{y \in A} u(y)}$, and

(2) for every other function u' satisfying condition (1) there is a constant k such that $u = k \cdot u'$.

Proof   First consider the existence part, (1), saying that the utility function u exists. Since it was assumed that the probability that x is chosen is nonzero, it also holds that $p(A \succ B) \neq 0$. It then follows saying that $p(x \succ A) = \dfrac{p(x \succ B)}{p(A \succ B)}$. We stipulate that $u(x) = k \cdot p(x \succ B)$, where $k > 0$. Then, since the elements of A are mutually exclusive, the axioms of the probability calculus guarantee that $p(x \succ A) = \dfrac{k \cdot p(x \succ B)}{\sum_{y \in A} k \cdot p(y \succ B)} = \dfrac{u(x)}{\sum_{y \in A} u(y)}$. We now move on to the uniqueness part, (2). Suppose that u' is another utility function defined as above and satisfying (1). Then, for every x in B, it holds that $u(x) = k \cdot p(x \succ B) = \dfrac{k \cdot u'(x)}{\sum_{y \in A} u'(y)}$.   □

The downside of the probabilistic theory is that it requires that $p \neq 0$, i.e. that each option is chosen with a nonzero probability. (This assumption is essential in the proof of Theorem 5.3, because no number can be divided by 0.) However, as pointed out above, the probability that you choose \$10 rather than \$20 if offered a choice between the two is likely to be 0. Call this the problem of perfect discrimination. The problem of perfect discrimination can be overcome by showing that for every set of objects B there exists some *incomparable* object x^* such that $p(x^* \succ x) \neq 0, 1$ for every x in B. Suppose, for example, that I wish to determine my utility of \$20, \$30 and \$40, respectively. In this case, the incomparable object can be taken to be a

Table 5.5

	u_1	u_2	u_3	u
$20	1/4	–	–	$\dfrac{1/4}{3/4} = 1/3$
$30	–	2/4	–	$\dfrac{2/4}{2/4} = 1$
$40	–	–	3/4	$\dfrac{3/4}{1/4} = 3$
photo	3/4	2/4	1/4	1

photo of my beloved cat Carla, who died when I was fourteen. The photo of this cat has no precise monetary value for me; my choice between money and the photo is always probabilistic. If offered a choice between $20 and the photo, the probability is 1/4 that I would choose the money; if offered a choice between $30 and the photo, the probability is 2/4 that I would choose the money; and if offered a choice between $40 and the photo, the probability is 3/4 that I would choose the money. This information is sufficient for constructing a single ratio scale for all four objects. Here is how to do it.

We use the three local scales as our point of departure. They have one common element; the photo of Carla. The utility of the photo is the same in all three pairwise choices. Let u(photo)=1. Then the utility of money is calculated by calibrating the three local scales such that u(photo) =1 in all of them. This is achieved by dividing the probability numbers listed above by 3/4, 2/4 and 1/4, respectively. Table 5.5 summarises the example. The symbols u_1–u_3 denote the three local scales and u denotes the single scale obtained by 'welding together' u_1–u_3.

Of course, there might exist some large amount of money that would make me choose the money over the photo with probability one. This indicates that the photo is not incomparable with every possible amount of money. However, this difficulty can be overcome by choosing some other beloved object to compare with, e.g. the only remaining photo of my daughter, or peace in the Middle East.

5.4 Can we define utility without being able to measure it?

Intuitively, it seems plausible to separate the meaning of the term utility from the problem of how to measure it. Consider the following analogy: We all know what it means to say that the mass of Jupiter exceeds that of Mars, but few of us are able to explain how to actually measure the mass of a planet. Therefore, if someone proposed a measurement procedure that conflicts with our intuitive understanding of mass, we would have reason not to accept that procedure as long as we wish to measure the thing we call mass. We are, under normal circumstances, not prepared to replace our intuitive concept of mass with some purely technical concept, even if the technical concept simplifies the measurement process. Does the analogous point apply to utility?

Theories of utility are sometimes interpreted as *operational definitions*, i.e. as definitions that fix the meaning of a term by setting up an empirical method for observing the entity the term is referring to. In this view, it does not make sense to distinguish meaning from measurement. As long as the concept of utility is merely used for descriptive purposes, i.e. for predicting and explaining choices, this operational approach seems fine. However, when we consider normative applications it is far from clear that an operational definition is what we are looking for. If we, for example, wish to say that a decision was rational *because* it maximised expected utility, it seems essential that the notion of utility we refer to is the true, *core notion* of utility. So what is this core notion of utility, with which operational procedures should be compared?

Philosophers taking a utilitarian approach to ethics frequently apply the notion of utility in moral contexts. These utilitarian philosophers often think of utility as a mental state. That my utility increases if I get a new car means that my mental state is transformed from one state into another, which is more valuable. Let us see if we can make any sense of this traditional utilitarian notion of utility. If we can, we will at least be able to say something interesting about the *meaning* of utility. To start with, it is helpful to divide the utility of an outcome or object into temporal intervals, such that the utility may vary from one interval to the next, but not within an interval. Call such intervals, which may be arbitrarily small, *moments* of utility. It is, of course, an empirical question whether moments of utility exist. It cannot be excluded that in some time periods, there are no constant moments of utility. To overcome this problem we assume that if m is an interval which cannot be divided into a sequence of constant intervals, then it is always possible to construct a constant interval m' covering the same time interval, such that $m \sim m'$, by choosing some m' having the right intensity.

A moment of utility is to be thought of as a property of an individual's experience within a certain time interval. The more an agent wants to experience a moment, the higher is the utility of the moment. Thus, the agent's well-informed preferences over different moments are likely to be the best way of determining the utility of moments. In this respect the utilitarian concept resembles von Neumann and Morgenstern's approach, since the latter also uses preferences for axiomatising utility.

Let $M = \{a, b, ...\}$ be a set of moments, and let $\succ$ be a relation on M representing strict preference. Indifference is represented by the relation $\sim$. We furthermore suppose that there is a binary operation $\circ$ on M. Intuitively, $a \circ b$ denotes the utility of first experiencing the utility moment a, immediately followed by the utility moment b. The set of utility moments is an *extensive structure* if and only if, for all $a, b, c, d \in M$, the axioms stated below hold true. We recognise the first two axioms from von Neumann and Morgenstern's theory, but note that they now deal with moments rather than lotteries.

Util 1 Either $a \succ b$ or $b \succ a$ or $a \sim b$

Util 2 If $a \succ b$ and $b \succ c$, then $a \succ c$

Util 3 $[a \circ (b \circ c)] \sim [(a \circ b) \circ c]$

Util 4 $a \succ b$ if and only if $(a \circ c) \succ (b \circ c)$ if and only if $(c \circ a) \succ (c \circ b)$

Util 5 If $a \succ b$, then there is a positive integer n such that $(na \circ c) \succ (nb \circ d)$, where na is defined inductively as $1a = a, (n + 1)a = (a \circ na)$.

The third axiom, Util 3, is mainly technical. It holds that it does not matter if b and c are attached to a or if c is attached to a and b. Util 4 states that in the case that a utility moment a is preferred to b, then $a \succ b$ even in the case that a moment c comes before or after those moments. Of course, this axiom does not imply that the entities that *cause* utility can be attached in this way. For example, if salmon is preferred over beef, it would be a mistake to conclude that salmon followed by ice cream is preferred to beef followed by ice cream. Moreover, Util 4 tells us that if the utility of eating salmon is preferred to the utility of eating beef, then the utility of eating salmon followed by the utility of eating ice cream after eating salmon is preferred to the utility of eating beef followed by the utility of eating ice cream after eating beef.

Util 5 is an Archimedean condition. It implies that even if d is very strongly preferred to c, then this difference can always be outweighed by a sufficiently large number of moments equal to a with respect to b, $a \succ b$, such that $(na \circ c) \succ (ny \circ d)$. This roughly corresponds to the Archimedean property of real numbers: if $b > a > 0$ there exists a finite integer n such that $na > b$, no matter how small a is. Util 5 is problematic if one thinks that there is some critical level of utility, such that a sequence of moments containing a sub-critical level moment should never be preferred to a sequence of moments not containing a sub-critical level moment. Personally I do not think there are any such critical levels, but to really argue for that point is beyond the scope of this book.

Theorem 5.4   If Util 1–5 hold for $\succ$ on a non-empty set of moments M and if $\circ$ is a binary operation on M, then there exists a real-valued function u on M such that

(1) $a \succ b$ if and only if $u(a) > u(b)$, and
(2) $u(a \circ b) = u(a) + u(b)$, and
(3) For every other function u' that satisfies properties (1) and (2) there exists some $k > 0$ such that $u' = ku$.

Theorem 5.4 follows from a standard theorem in measurement theory. (See e.g. Krantz *et al.* 1971.) I shall spare the reader from the proof. However, note that the axioms listed above do not say anything about what is being measured. It is generally agreed that they hold for mass and length, and if the (hedonistic) utilitarians are right they also hold for moments of utility. However, if they do hold for moments of utility they merely fix the meaning of the concept. The axioms say almost nothing about how utility could be measured in practice. Are agents really able to state preferences not between, say, salmon and beef, but between the mental states caused by having salmon or beef, respectively? And are they really able to do so even if the comparison is made between hypothetical mental states which are never experienced by anyone, as required by the theory?

Exercises

5.1 Which preference ordering is represented by the following ordinal utility function: $u(a) = 7$, $u(b) = 3$, $u(c) = 34$, $u(d) = -430$, $u(e) = 3.76$?
5.2 The following function u is not an ordinal utility function. Why not? $u(a) = 7$, $u(b) = 3$, $u(c) = 34$, $u(d) = -430$, $u(e) = 3.76$, $u(e) = 12$.

5.3 Your preferences are transitive and asymmetric, and you prefer a to b and b to c. Explain why it has to be the case that you do not prefer c to a.

5.4 Strict preferences are irreflexive, meaning that no x is strictly preferred to itself. Show that asymmetry implies irreflexivity.

5.5 Show that negative transitivity is logically equivalent with the following claim: $x \succ z$ implies that, for all y in B, $x \succ y$ or $y \succ z$.

5.6 Show that if $\succ$ is asymmetric and negatively transitive, then $\succ$ is transitive.

5.7 You are indifferent between receiving A for sure and a lottery that gives you B with a probability of 0.9 and C with a probability of 0.1. You are also indifferent between receiving A for sure and a lottery that gives you B with a probability of 0.6 and D with a probability of 0.4. All of your preferences satisfy the von Neumann–Morgenstern axioms.
   (a) What do you prefer most, C or D?
   (b) Calculate the (relative) difference in utility between B and C, and between B and D.
   (c) If we stipulate that your utility of B is 1 and your utility of C is 0, what are then your utilities of A and D?

5.8 The continuity axiom employed by von Neumann and Morgenstern holds that if $A \succ B \succ C$ then there exist some probabilities p and q such that $ApC \succ B \succ AqC$. Let $A = \$10,000,001$ and $B = \$10,000,000$, and $C = 50$ years in prison. (a) Do you think it is really true that there are *any* values of p and q such that $ApC \succ B \succ AqC$ truly describes your preferences? (b) Psychological studies suggest that most people cannot distinguish between very small probabilities, i.e. that their preferences over lotteries in which there is a small probability of a very bad outcome is unaffected by exactly how small the probability is. Does this show that there is something wrong with von Neumann and Morgenstern's theory?

5.9 You prefer a fifty-fifty chance of winning either $100 or $10 to a lottery in which you win $200 with a probability of 1/4, $50 with a probability of 1/4, and $10 with a probability of 1/2. You also prefer a fifty-fifty chance of winning either $200 or $50 to receiving $100 for sure. Are your preferences consistent with von Neumann and Morgenstern's axioms?

5.10 You somehow know that the probability is 75% that your parents will complain about the mess in your room the next time they see you. What is their utility of complaining?

5.11 You somehow know that the probability is 5% that you will tidy up your room before your parents come and visit you. What is best for you, to tidy up your room or live in a mess?

5.12 The conclusion of Exercise 5.11 may be a bit surprising – can you really find out what is best for you by merely considering what you are likely to do? For example, the probability is 0.99 that a smoker will smoke another cigarette, but it seems false to conclude that it would be better for the smoker to smoke yet another cigarette. What could the advocate of the probabilistic theory say in response to this objection?

Solutions

5.1 $c \succ a \succ e \succ b \succ d$

5.2 Because u assigns two different numbers to one and the same argument: $u(e) = 3.76$ and $u(e) = 12$.

5.3 It follows from transitivity that you prefer a to c, and by applying asymmetry to that preference we find that it is not the case that c is preferred to a.

5.4 Substitute y for x in the definition of asymmetry.

5.5 By contraposition, the rightmost part of the statement can be transformed into the following logically equivalent statement: not ($x \succ y$ or $y \succ z$ for all y in B) implies not $x \succ z$. This is equivalent with: for all y in B, not $x \succ y$ and not $y \succ z$ implies not $x \succ z$. This is negative transitivity.

5.6 Suppose that $x \succ y$ and $y \succ z$. From asymmetry and Exercise 5.5 we can conclude that: $x \succ y$ implies that either $z \succ x$ or $x \succ x$. The first possibility, $y \succ z$, is inconsistent with what we initially suppose and can hence be ruled out. Hence, $x \succ z$, which gives us transitivity. Asymmetry implies that not $y \succ x$ and not $z \succ y$. By applying negative transitivity to these preferences we get: not $z \succ x$. From completeness it follows that either $x \succ z$ or $x \sim z$.

5.7 (a) D (b) The difference between B and C is exactly four times the difference between B and D. (c) $u(A) = 90$ and $u(D) = 75$.

5.8 (a) No matter how small the probabilities p and q are it seems better to take B for sure. Why risk everything for one extra dollar? (b) The psychological evidence suggests that the von Neumann–Morgenstern theory does not accurately describe how people do in fact behave.

Whether it is a valid normative hypothesis depends on what one thinks about the answer to (a).

5.9   No. Your preferences violate the independence axiom.

5.10 Their utility of complaining is three times that of not complaining.

5.11 It is better for you to leave your room as it is; the utility of that option is 19 times that of tidying up the room.

5.12 In this context 'better for' means 'better as viewed from the decision maker's present viewpoint and given her present beliefs and desires'. The point that the smoker may in fact die of lung cancer at some point in the future is perfectly consistent with this notion of 'better'.

6   The mathematics of probability

The following decision problem appeared some years ago in Marilyn vos Savant's Ask Marilyn column in the *Parade Magazine*.

> You are a contestant in a game show hosted by Monty Hall. You have to choose one of three doors, and you win whatever is behind the door you pick. Behind one door is a brand new car. Behind the other two is a goat. Monty Hall, who knows what is behind the doors, now explains the rules of the game: "First you pick a door without opening it. Then I open one of the other doors. I will always pick a door to reveal a goat. After I have shown you the goat, I will give you the opportunity to make your final decision whether to stick with your initial door, or to switch to the remaining door." Should you accept Monty Hall's offer to switch?

The Monty Hall problem has a simple and undisputable solution, which all decision theorists agree on. If you prefer to win a brand new car rather than a goat, then you ought to switch. It is irrational not to switch. Yet, a considerable number of Marilyn vos Savant's readers argued that *it makes no difference* whether you switch or not, because the probability of winning the car must be 1/3 no matter how you behave. Although intuitively plausible, this conclusion is false.

The point is that Monty Hall's behaviour actually reveals some extra information that makes it easier for you to locate the car. To see this, it is helpful to distinguish between two cases. In the first case you initially pick one of the goats. Then, since there is only one goat left, Monty Hall *must* open the door with the second goat. That is, there is only one door he *can* open. Hence, given that you initially pick a goat, Monty Hall will indirectly reveal where the car is; in those cases you should certainly switch. In the second case you initially pick the car. In that case it would of course be bad to switch. However, although you do not know whether your initial pick is a goat or the car, you know that the *probability* that you

Table 6.1

	1/3	1/3	1/3	
Contestant chooses:	Goat 1	Goat 2	Car	
	1	1	1/2	1/2
Host reveals:	Goat 2	Goat 1	Goat 2	Goat 1
Total probability:	1/3	1/3	1/6	1/6
Stay:	Goat 1	Goat 2	Car	Car
Switch:	Car	Car	Goat 1	Goat 2

have initially picked a goat is 2/3. Hence, since it is *certain* that you will get the car if you initially pick a door with a goat and then switch, you will on average win the car two times out of three if you decide to always switch. However, if you always refuse the offer to switch, you will on average win the car only one time out of three (that is, if and only if you picked the car in your initial choice). See Table 6.1.

The moral of the Monty Hall problem is: Anyone who wishes to make rational decisions needs to reason correctly about probabilities, because probability theory is one of the most important tools in decision theory. This chapter gives an overview of the probability calculus, i.e. the mathematical theory of probability. The next chapter focuses on philosophical problems related to the probability calculus.

6.1 The probability calculus

Imagine that you roll a traditional and fair die twice. There are exactly 36 possible outcomes: You get 1 in the first roll and 1 in the second, or 1 in the first and 2 in the second, or 2 in the first and 1 in the second, and so on and so forth. In Table 6.2 each outcome is represented by two numbers $\langle i, j \rangle$ – mathematicians call this an 'ordered pair' – where i refers to the outcome of the first roll and j to the second roll. So, for example, $\langle 1, 4 \rangle$ and $\langle 4, 1 \rangle$ are two different outcomes.

Before we proceed, it is helpful to introduce some terminology. Let S be the set of all possible outcomes of the *random experiment* explained above.

Table 6.2

$\langle 1,1 \rangle$	$\langle 1,2 \rangle$	$\langle 1,3 \rangle$	$\langle 1,4 \rangle$	$\langle 1,5 \rangle$	$\langle 1,6 \rangle$
$\langle 2,1 \rangle$	$\langle 2,2 \rangle$	$\langle 2,3 \rangle$	$\langle 2,4 \rangle$	$\langle 2,5 \rangle$	$\langle 2,6 \rangle$
$\langle 3,1 \rangle$	$\langle 3,2 \rangle$	$\langle 3,3 \rangle$	$\langle 3,4 \rangle$	$\langle 3,5 \rangle$	$\langle 3,6 \rangle$
$\langle 4,1 \rangle$	$\langle 4,2 \rangle$	$\langle 4,3 \rangle$	$\langle 4,4 \rangle$	$\langle 4,5 \rangle$	$\langle 4,6 \rangle$
$\langle 5,1 \rangle$	$\langle 5,2 \rangle$	$\langle 5,3 \rangle$	$\langle 5,4 \rangle$	$\langle 5,5 \rangle$	$\langle 5,6 \rangle$
$\langle 6,1 \rangle$	$\langle 6,2 \rangle$	$\langle 6,3 \rangle$	$\langle 6,4 \rangle$	$\langle 6,5 \rangle$	$\langle 6,6 \rangle$

S is called the *sample space*. In our example S is finite, but one can easily imagine cases in which S is infinite. For example, if we roll a die until a six turns up and take the number of rolls needed for getting a six to be the outcome of a random experiment, then the sample space will be $S = \{1, 2, 3, ...\}$. This set is countably infinite.

To keep the mathematics as simple as possible we assume that the sample space is either finite or countably infinite. This is a significant limitation, because it implies that we will not be able to answer questions like: "What is the probability of choosing the real number r from the interval $1 \geq r \geq 0$?" As you might recall from high school, every interval of real numbers is uncountably infinite. The concepts needed for analysing such problems are introduced in more advanced books on probability theory.

An *event* A is a subset of the sample space. Hence, to first roll a six and then roll a five, $\langle 6, 5 \rangle$, is an event. To roll $\langle 6, 5 \rangle$ *or* $\langle 5, 6 \rangle$ *or* $\langle 1, 1 \rangle$ is also an event, because $\langle 6, 5 \rangle$ *or* $\langle 5, 6 \rangle$ *or* $\langle 1, 1 \rangle$ is a subset of S. However, to just roll a six and never finish the random experiment is not an event, because just rolling a six is not a subset of S. By definition, if A and B are events, so is A-or-B and A-and-B and not-A and not-B. Furthermore, for every event A, there is a complement of A, denoted A^c. The complement of A is the set of all events that are in S but not in A. Therefore, the complement of $\langle 1, 1 \rangle$ is the set of all events that can be constructed from every outcome except $\langle 1, 1 \rangle$ in the matrix above. Since S is the set of all possible outcomes, $S^c = \varnothing$, i.e. the complement of S is the empty set.

Probabilities are represented by real numbers between 0 and 1 (including the end points). We say that a function p is a *probability measure* only if (but not if and only if) p assigns a number $1 \geq p(A) \geq 0$ to every event A in S. This means that p is a function that takes an event as its argument and then returns a number between 0 and 1 that represents the probability of that event. Some reflection reveals that in our example, $p(\langle 6,5 \rangle) = \dfrac{1}{36}$ and $p(\langle 6,\ 5 \rangle \ or\ \langle 5,\ 6 \rangle \ or\ \langle 1,\ 1 \rangle) = \dfrac{3}{36}$.

The probability calculus is an axiomatised theory. This means that all true (mathematical) statements about probability can be derived from a small set of basic principles, or axioms. Consider the following axioms, first proposed by Kolmogorov in 1933.

(1) Every probability is a real number between 0 and 1.
(2) The probability of the entire sample space is 1.
(3) If two events are mutually exclusive, then the probability that one of them will occur equals the probability of the first plus the probability of the second.

Naturally, these three axioms can also be stated in a mathematical vocabulary.

Kolmogorov 1 $1 \geq p(A) \geq 0$
Kolmogorov 2 $p(S) = 1$
Kolmogorov 3   If $A \cap B = \emptyset$, then $p(A \cup B) = p(A) + p(B)$

The first axiom articulates the idea that the probability of every event lies between 0 and 1, as explained above. The second axiom states that the probability of the entire sample space is 1; hence, the probability is 1 that the outcome of rolling a die twice will be one of those listed in Table 6.2. In order to come to grips with the third axiom we need to understand the set-theoretic operations of intersection and union: $A \cap B$ is the intersection of A and B, that is, $A \cap B$ contains only those elements that are both in A and in B. Hence, to say that $A \cap B = \emptyset$ just means that A and B have no common elements – they are mutually exclusive. The formula $A \cup B$ denotes the union of A and B, that is, all elements that are either in A or in B or in both. Hence, the third axiom articulates the claim that if two events are mutually exclusive, then the probability that one of them will occur equals the probability of the first plus the probability of the second. In the die example, all 36 (non-composite) outcomes are mutually exclusive, because only one of them can occur every time you roll a die twice. However, $\langle 6,\ 5 \rangle$ and $\langle 6,\ 5 \rangle \cup \langle 5,\ 6 \rangle$ are not mutually exclusive, because $\langle 6,\ 5 \rangle$ is an element in both events.

Up to this point we have formulated the theory of probability in set-theoretical terms. This is the traditional way of presenting the probability calculus, preferred by mathematicians. However, some philosophers prefer to present the probability calculus in the language of propositional logic, since this entails that the usual logical operations and inference rules can be directly applied to probabilities. In what follows we shall not take a stand on which approach is best. Let us just note that the two ways of presenting the theory are closely interrelated. For example, the proposition A = "I roll a die and get a six, and then roll it again and get a five" is true if and only if event $\langle 6, 5 \rangle$ occurs. In order to spell out the propositional approach in more detail, we stipulate that a pair of propositions is mutually exclusive if and only if it holds that if one is true, then the other must be false. Now, by taking A and B to be arbitrary propositions (instead of events), Kolmogorov's axioms can be reformulated in the following way.

Kolmogorov 1* $1 \geq p(A) \geq 0$

Kolmogorov 2*   If A is a logical truth, then $p(A) = 1$

Kolmogorov 3*   If A and B are mutually exclusive, then $p(A \lor B) = p(A) + p(B)$

The difference between the two approaches is important, since they assign probabilities to different *kinds of things*. If we use the first set of axioms we assign probabilities to events, that is, things that take place in the external world. The second set assigns probabilities to linguistic entities, viz. propositions. Propositions are the kind of thing we use when we reason and make inferences. If we believe that probability calculus is part of some more general theory of valid inferences and rational reasoning, then it arguably makes more sense to base it on the propositional approach. Or this is at least the majority view among philosophers in the contemporary literature.

The probability axioms have a number of interesting implications. To start with, note that

Theorem 6.1 $p(A) + p(\neg A) = 1$

Proof $A \lor \neg A$ is a logical truth; therefore, by axiom 2* we have $p(A \lor \neg A) = 1$. Furthermore, A and $\neg A$ are mutually exclusive, so by axiom 3* we have $p(A \lor \neg A) = p(A) + p(\neg A)$. Now, by combining these two equalities we get $p(A) + p(\neg A) = p(A \lor \neg A) = 1$.   □

Theorem 6.2   If A and B are logically equivalent, then $p(A) = p(B)$

Proof   If A and B are logically equivalent, then A and $\neg B$ are mutually exclusive (because both propositions cannot be true simultaneously).

From axiom 3* we have $p(A \lor \neg B) = p(A) + p(\neg B)$. Furthermore, $A \lor \neg B$ is certain to be true for logical reasons, so by axiom 2* we have $p(A \lor \neg B) = 1$. Hence,

$$p(A) + p(\neg B) = 1 \tag{1}$$

Now we use Theorem 6.1, i.e. the fact that

$$p(B) + p(\neg B) = 1 \tag{2}$$

Since $p(\neg B)$ has the same value in (1) and (2) it follows that $p(A) = p(B)$.   □

Theorem 6.3

(i) $$p(A \lor B) = p(A) + p(B) - p(A \land B)$$

and

(ii) $$p(A \to B) = p(\neg A) + p(B) - p(\neg A \land B4)$$

Proof   Part (i) $A \lor B$ and $(\neg A \land B) \lor (A \land \neg B) \lor (A \land B)$ are logically equivalent, so Theorem 6.2 guarantees that the two formulas have the same probability. However, $(\neg A \land B)$ and $(A \land \neg B) \lor (A \land B)$ are mutually exclusive, so Axiom 3* implies that

$$\begin{aligned} p(A \lor B) &= p\big[(\neg A \land B) \lor (A \land \neg B) \lor (A \land B)\big] \\ &= p(\neg A \land B) + p\big[(A \land \neg B) \lor (A \land B)\big] \end{aligned} \tag{3}$$

The rightmost term of this equality, $(A \land \neg B) \lor (A \land B)$, is equivalent with A, so

$$p(A \lor B) = p(\neg A \land B) + p(A) \tag{4}$$

Now take a look at Theorem 6.3(i): in some way or another we have to get rid of $p(\neg A \land B)$ and introduce $p(B)$ and $p(A \land B)$. Let's see what happens if we add $p(A \land B)$ to both sides of (4).

$$p(A \lor B) + p(A \land B) = p(\neg A \land B) + p(A) + p(A \land B) \tag{5}$$

We now wish to clean up the right-hand side of (5). Since $(\neg A \land B) \land (A \land B)$ is equivalent with B, we get

$$p(A \lor B) + p(A \land B) = p(A) + p(B) \tag{6}$$

Hence, $p(A \lor B) = p(A) + p(B) - p(A \land B)$.

Part (ii) $A \rightarrow B$ and $\neg A \vee B$ are logically equivalent, so all we have to do is to substitute A for $\neg A$ in Part (i).   □

Example (i)   What is the probability of getting rich and happy? You know that the probability of getting rich is 0.2 and the probability of getting happy is 0.1, whereas the probability of getting either rich or happy is 0.25.

Solution   Theorem 6.3(i) tells us that: $p(A \wedge B) = p(A) + p(B) - p(A \vee B)$. Hence, the probability of getting rich and happy is rather small. In fact, it is only 0.2 + 0.1 − 0.25 = 0.05.

Example (ii)   You roll four fair dice. What is the probability of getting at least one six?

Solution   Let A mean that you get at least one six in a roll of four dice. Then, $\neg A$ is logically equivalent with getting no six at all when rolling four dice. We know that $p(\neg A) = \frac{5}{6} \cdot \frac{5}{6} \cdot \frac{5}{6} \cdot \frac{5}{6}$, so by Theorem 6.1 we have $p(A) = 1 - p(\neg A) = 1 - \frac{5}{6} \cdot \frac{5}{6} \cdot \frac{5}{6} \cdot \frac{5}{6} \approx 0.52$.

6.2 Conditional probability

Many statements about probability are conditional. For example, a decision maker visiting a posh restaurant may wish to determine the probability that the fish on the menu is fresh *given that* it is Monday (in most restaurants this conditional probability is low), or the probability that the beef will be rare *given that* one asks the chef for a medium cooked piece of meat. In order to calculate conditional probabilities, we first need to define the concept. Consider the following definition of what it means to talk about 'the probability of A given B', abbreviated as $p(A|B)$. We assume that $p(B) \neq 0$.

Definition 6.1 $p(A|B) = \dfrac{p(A \wedge B)}{p(B)}$

Example (iii)   You roll a fair die twice. Given that the first roll is a 5, what is the probability that the total sum will exceed 9? It takes little reflection to see that the answer is $\frac{1}{3}$, because the second time you roll the die you have to get either a 5 or 6. Fortunately, this intuitive calculation is consistent with the result we get if we apply Definition 6.1. Let A = 'the sum exceeds 9' (i.e. is

10, 11 or 12), and let B = 'the first roll is a 5'. By looking at Table 6.2, we see that there are exactly two mutually exclusive outcomes in which it is true that the total exceeds 9 and the first roll is a 5, viz. $\langle 5, 5\rangle$ and $\langle 5, 6\rangle$. Hence, $p(A \wedge B) = \dfrac{2}{36}$. By looking at Table 6.2 again, we find that $p(B) = \dfrac{6}{36}$. Hence,

$$p(A|B) = \frac{p\left(\dfrac{2}{36}\right)}{p\left(\dfrac{6}{36}\right)} = \frac{2}{6} = \frac{1}{3}.$$

When you roll two dice we tacitly assume that the outcome of the second roll does not depend on the outcome of the first roll. We simply take for granted that the outcomes of the two rolls are independent. It is, of course, desirable to render our intuitive understanding of independence more precise, and this can be achieved by observing that the concept of independence is closely related to the concept of conditional probability. If the probability of rain tomorrow is independent of whether the republican candidate wins the next presidential election, then it holds that $p(\text{rain}) = p(\text{rain}|\text{republican})$. More generally speaking, it holds that

Definition 6.2 A is independent of B if and only if $p(A) = p(A|B)$.

Definitions 6.1 and 6.2 can be applied for calculating the probability of many conjunctions. Consider the following theorem.

Theorem 6.4   If A is independent of B, then $p(A \wedge B) = p(A) \cdot p(B)$.

Proof   By combining Definitions 6.1 and 6.2 we see that $p(A) = \dfrac{p(A \wedge B)}{p(B)}$, which gives us Theorem 6.4.   □

Example (iv)   You plan to rob four banks and then escape to Mexico. In each robbery the probability of getting caught is $\dfrac{1}{3}$, and the outcome of each robbery is independent of that of the others. What is the probability that you end up in jail?

Solution   The probability of ending up in jail is $1 - p(\text{never getting caught})$. Since the outcomes of the four robberies are independent, the probability of never getting caught is $\left(1 - \dfrac{1}{3}\right)^4 = \dfrac{2^4}{3^4} = \dfrac{16}{81}$. Hence, the probability of ending up in jail is quite high: $1 - \dfrac{16}{81} \approx 0.8$.

6.3 Bayes' theorem

Bayes' theorem is the most important and widely known theorem presented in this chapter. Briefly put, the theorem shows how conditional probabilities can be 'reversed', which is of great importance in many applications of probability theory. Imagine, for instance, that an engineer has observed a welding crack in a vital part of a gearbox. The engineer wishes to calculate the probability that the gearbox will break down (B) given the appearance of a welding crack (A). Statistics show that 90% of all broken gearboxes (of the relevant type) have welding cracks. Hence, $p(A|B) = 0.9$. Furthermore, the engineer knows that 10% of all gearboxes break down during their predicted lifespan, and that 20% of all gearboxes have welding cracks. Hence, $p(B) = 0.1$ and $p(A) = 0.2$. Now, the engineer can solve the problem by inserting the numbers into the following very simple version of Bayes' theorem, which is sometimes called *the inverse probability law*.

Theorem 6.5 $\quad p(B|A) = \dfrac{p(B) \cdot p(A|B)}{p(A)}$ given that $p(A) \neq 0$

Proof According to the definition of conditional probability, $p(A|B) = \dfrac{p(A \wedge B)}{p(B)}$, and $p(B|A) = \dfrac{p(B \wedge A)}{p(A)}$. Since $A \wedge B$ and $B \wedge A$ are logically equivalent, Theorem 6.2 entails that $p(B) \cdot p(A|B) = p(A \wedge B) = p(B \wedge A) = p(A) \cdot p(B|A)$. Hence, $p(B|A) = \dfrac{p(B) \cdot p(A|B)}{p(A)}$. $\qquad\square$

By inserting the numbers listed above into Theorem 6.5, the engineers find that the probability that the gearbox will break down given the appearance of a welding crack is quite high:

$$p(B|A) = \frac{0.1 \cdot 0.9}{0.2} = 0.45 \qquad\qquad (7)$$

In many cases it is difficult to determine what the unconditional probabilities of A and B are. In the welding crack example, it was simply taken for granted that those unconditional probabilities were known. However, by playing with Theorem 6.5 a bit, we can eliminate at least one of the unconditional probabilities that appear in the formula, viz. the percentage of all gearboxes that have welding cracks, $p(A)$. Consider the following version of Bayes' theorem, which is probably the most well-known formulation.

Theorem 6.6 $p(B|A) = \dfrac{p(B) \cdot p(A|B)}{[p(B) \cdot p(A|B)] + [p(\neg B) \cdot p(A|\neg B)]}$ given that $p(A) \neq 0$.

Proof We use Theorem 6.5 as our point of departure. Hence, it suffices to show that $p(A) = p(B) \cdot p(A|B) + p(\neg B) \cdot p(A|\neg B)$. To start with, note that A and $(A \wedge B) \vee (A \wedge \neg B)$ are logically equivalent, and that the disjuncts of the latter formula are mutually exclusive. Hence, by Axiom 3* and Theorem 6.2, $p(A) = p(A \wedge B) + p(A \wedge \neg B)$. The theorem then follows by inserting the two terms on the right-hand side into the definition of conditional probability. To see this, recall that $p(A|B) = \dfrac{p(A \wedge B)}{p(B)}$. It follows that $p(B) \cdot (A|B) = p(A \wedge B)$. Hence, $p(A) = p(B) \cdot p(A|B) + p(\neg B) \cdot p(A|\neg B)$. □

Example (v) Your doctor suspects that you may be suffering from Wilson's disease, a rare disease that affects about 1 in 30,000. Given that you have the disease, the test offered by the doctor will show positive with probability 0.995. Unfortunately, the test will also show positive with probability 0.01 when applied to a healthy person. What is the probability that you have Wilson's disease given that the test shows positive?

Solution By inserting the numbers into Bayes' theorem we find that the probability that you suffer from Wilson's disease is surprisingly low:

$$p(\text{disease}|\text{positive}) = \frac{\dfrac{1}{30,000} \cdot 0.995}{\left[\dfrac{1}{30,000} \cdot 0.995\right] + \left[\left(1 - \dfrac{1}{30,000}\right) \cdot 0.01\right]} \approx 0.0033 \quad (8)$$

Example (vi) It is Christmas Day and you are sailing in the North Atlantic. You observe that the barometer is falling more than 1 mb per hour, so you are a bit worried that a storm is coming. According to the best available meteorological data, the probability that the barometer falls more than 1 mb per hour given that a storm is coming is 0.95. The probability that it falls more than 1 mb per hour given that a storm is not coming, however, is just 1 in 1,000. Furthermore, the probability that there is a storm in the North Atlantic any given day in December is 0.25. What is the probability that a storm is coming given that the barometer falls more than 1 mb per hour?

Solution You should definitely prepare the ship for storm, because

$$p(\text{storm}|\text{falls}) = \frac{0.25 \cdot 0.95}{[0.25 \cdot 0.95] + [(1 - 0.25) \cdot 0.001]} \approx 0.997 \quad (9)$$

The probability $p(B)$, i.e. the unconditional probability that there will be a storm in the North Atlantic or that the patient is suffering from Wilson's disease, is called the *prior* probability. In many cases it is difficult or impossible to know the prior probability of an event. For example, when AIDS was discovered in San Francisco in the 1980s no one knew what the prior probability of this new disease was. However, as soon as the number of patients suffering from the disease had been counted it was easy to calculate the prior probability. That said, the prior can of course vary over time. Today the unconditional probability that a randomly chosen person in San Francisco will be HIV positive is fairly low. Sadly, the prior is much higher in some parts of southern Africa.

6.4 The problem of unknown priors

It is easy to imagine examples in which it is very difficult or even impossible to know the prior probability. What was, for example, the prior probability that the first human space flight (by Soviet cosmonaut Yuri Gagarin in 1961) would be successful? Or what is the prior probability that the republican candidate will win the next election? The fact that prior probabilities are often difficult to determine has given rise to a number of interesting problems, which are widely discussed by statisticians. At the heart of this discussion lies an interesting proposal for how Bayes' theorem could be applied even when the prior probability is unknown. Let us take a closer look at this proposal.

Imagine that you are in Las Vegas on vacation. After drinking a glass of complimentary champagne in your suite on the 29th floor, you feel ready to visit the casino and decide to play roulette. In American roulette the wheel has 38 numbered pockets marked 0, 00, 1, 2, 3, ..., 36. The house wins whenever the ball falls into the pockets marked 0 or 00, so the probability for this should be 1 in 19. However, the first five times you play, the house wins every single time. You therefore start to suspect that the wheel is biased, and after a while you recall a newspaper article saying that a number of roulette wheels in Las Vegas – no one knows how many – have been manipulated such that the house wins on average one out of two times. Now, what is the probability that the roulette wheel in front of you is one of the manipulated ones?

Let us try to solve this problem by applying Bayes' theorem. Let $p(B|5H)$ denote the probability that the wheel is biased given that the house wins five times in a row. Since all five trials are independent of each other $p(5H|\neg B) = \left(\frac{1}{19}\right)^5$, and according to the newspaper article $p(5H|B) = \left(\frac{1}{2}\right)^5$. Note that the second probability is *much* higher than the first, so your observations seem to fit better with the hypothesis that the wheel is biased. However, the problem is that you do not know the prior probability of facing a biased wheel, that is, the unconditional probability $p(B)$. All you know is that *some* wheels have been manipulated. Hence, when you insert your information into Bayes' theorem you will face an equation with two unknown variables, $p(B|5H)$ and $p(B)$.

$$p(B|5H) = \frac{p(B) \cdot p\left(\frac{1}{2}\right)^5}{\left[p(B) \cdot p\left(\frac{1}{2}\right)^5\right] + \left[(1 - p(B)) \cdot p\left(\frac{1}{19}\right)^5\right]} \tag{10}$$

In order to solve this problem we shall introduce a clever trick. We just ask you to *choose whatever value of $p(B)$ you wish*. If you believe that there are many manipulated roulette wheels in Las Vegas you choose a high prior, say 0.6, but if you trust the casino and feel that the owner respects the law, then you choose a much lower prior. Let us suppose that you trust the casino and choose a prior of 0.01. That is, before starting to play your unconditional probability that the wheel is biased is 0.01. Now, by inserting this number into Bayes' theorem we find that $p(B|5H) = 0.9987$. Furthermore, we can easily calculate your conditional probability that the wheel is biased given your observations, that is, $p(\neg B|5H)$. It is 0.0013. So after having observed five spins, each being independent of the others, you have changed your view dramatically. More precisely put, by applying Bayes' theorem you have been able to *update* your initial beliefs in light of new information. The new probability that the wheel is biased is called the *posterior* probability. Table 6.3 summarises what you believe about the roulette wheel before and after having watched the house win five times in a row.

Table 6.3

Before ('prior probability')	After ('posterior probability')
Wheel is biased: $p(B) = 0.01$	Wheel is biased: $p(B|5H) = 0.9987$
Wheel is not biased: $p(\neg B) = 0.99$	Wheel is not biased: $p(\neg B|5H) = 0.0013$

Is there anything wrong with this way of reasoning? Recall that your prior probability, $p(B) = 0.01$, was just a mere guess. What if $p(B) = 0.001$? Or if $p(B) = 0.5$? Well, by inserting these number into the formula we find that $p(B|5H) = 0.9872$ and $p(B|5H) = 0.9999$, respectively. Both these numbers are very close to 0.9987, so in this particular example it does not really matter which of those three priors you choose.

But what if you initially thought that only 1 in 10,000 wheels were manipulated, i.e. what if $p(B) = 0.0001$? Then your posterior probability that the wheel is biased would be 'only' 0.8856, which is much less than 0.9999. This seems to indicate that it is, after all, important to choose a prior that is close to the 'correct' one. However, in response to this remark many statisticians invoke a second trick. These statisticians argue that a rational decision maker should simply ask the croupier to spin the wheel a couple of more times, and then apply Bayes' theorem again to the new data set. This time the 'old' posterior probability of 0.8856 should be used as the 'new' prior. Or, put in a slightly different way, the idea is to apply Bayes' theorem a second time, in which one starts off from the value one got from applying the theorem the first time. For instance, if the house were to win five times in a row the second time (which is of course very unlikely!) and if $p(B) = 0.8856$, it turns out that the new posterior probability is $p(B|5H) = 0.9999$. So even if your initial prior were extremely low, you would nevertheless come to believe that the wheel is biased. This is because you successively acquire new data that you use to 'wash out' incorrect prior probabilities.

It is important to understand this idea of how incorrect priors can be 'washed out'. However, to characterise the 'washing-mechanism' in a technically precise way would require too much mathematics. All I can offer in this book is a non-technical sketch. Simply put, incorrect priors are washed out by applying Bayes' theorem over and over again. Each time Bayes' theorem is applied for updating the probability that the wheel is biased, one will get somewhat closer to the truth. The reason why one will always get closer to the truth is that Bayes' theorem takes into account both the prior probability as well as the observed outcome of some random experiment. The new posterior will always lie somewhere between those two values. Note that in the example above, I assume that you observed five rounds of roulette at a time. This was a superfluous assumption, made for illustrative purposes only. Nothing excludes that you apply Bayes' theorem after every single round. Hence, even after the house has won just once your posterior probability that the wheel is biased will increase, given that your prior is low. (Is this not odd? How can a single spin of the roulette wheel be allowed to affect your confidence that it is biased!?) However, the more times you apply Bayes' theorem, the closer you will get to the truth – your posterior probability will converge towards some value, and it will converge towards the *same* value no matter what your initial prior was! As I said, all this can be proved in a mathematically satisfactory way, but this is not the right occasion for this.

It should also be pointed out that convergence results can be applied for resolving disagreements about probabilities in a rational way. If two or more decision makers initially disagree about the unconditional probability of some random event, i.e. if they start off with different priors, then this disagreement could be resolved by applying Bayes' theorem over and over again every time they receive new data. That is, given that the random experiment is repeated sufficiently many times, they will eventually come to agree on the probability of the event they are interested in. Therefore, it does not really matter if a group of people do not know what the 'true' prior is; no matter what numbers they start off with they will always come to agree in the long run.

Exercises

6.1 A card deck contains 52 cards, half of which are red and half of which are black. You randomly draw a card, put it back, and then draw a second card.
   (a) What is the probability of drawing two red cards?
   (b) What is the probability of drawing two red or two black cards?
   (c) What is the probability of drawing one red and one black card?

6.2 You draw two cards from a card deck, without putting back the first before drawing the second.
   (a) What is the probability of drawing two red cards?
   (b) What is the probability of drawing two red or two black cards?
   (c) What is the probability of drawing one red and one black card?
   (d) What is the probability that the second card is red given that the first is red?
   (e) What is the probability that the second card is red given that the first is black?

6.3 The probability that the next president will be a democrat is 1/2, and the probability that the next president will be a woman is 0.3, but the probability that the next president will be a woman and a democrat is just 0.1. What is the probability that the next president will be democrat or a woman?

6.4 You know the following: $p(\neg A)$ is 0.6, and $p(B)$ is 0.3, and $p(A \rightarrow B)$ is 0.7. What is $p(\neg A \wedge B)$?

6.5 You roll a fair die twice. What is the probability that the total sum will exceed 5, given that the first roll is 1?

6.6 Two fair six-sided dice are rolled. (a) What is the probability that their sum is 7? (b) Does the probability that the sum is 7 depend on the score shown on the first die?

6.7 You know the following: $p(A) = 0.1$, and $p(B|A) = 1$, and $p(B|\neg A) = 0.3$. What is $p(A|B)$?

6.8 According to an article in a computer magazine 5.7% of all computers break down during their predicted lifespan, and 9.5% of all computers are infected by a virus. Furthermore, 92% of all computers that break down are infected by a virus. What is the probability that a computer that is infected by a virus breaks down?

6.9 Your doctor suspects that you may be suffering from a rare disease that affects about 1 in 50,000. Given that you have the disease, the test offered by the doctor will show positive with probability 0.9. Unfortunately, the test will also show positive with probability 0.01 when applied to a healthy person. What is the probability that you have the disease given that the test is positive? What is your opinion about the test offered by the doctor?

6.10 You toss a coin three times, and it lands heads up every time. You're a-priori probability that the coin is biased to land heads is 0.1. What is your a-posteriori probability that the coin is biased to always land heads up?

6.11 Your a-priori probability that the coin is biased is 0. Prove that no matter how many times you toss it and update the probability in light of the information received, the probability can never become nonzero.

6.12 Prove that if A is independent of A, then $p(A) = 0$ or $p(A) = 1$.

6.13 What is the probability that at least two children in a group of n children, all born in 2009, share a birthday? (Why did I assume that all children were born in 2009?)

6.14 Rumour has it that the following question was asked many years ago to students applying to study philosophy at the University of Cambridge: "Four witnesses, A, B , C and D, at a trial each speak the truth with probability $\frac{1}{3}$ independently of each other. In their testimonies, A claimed that B denied that C declared that D lied. What is the conditional probability that D told the truth?"

Solutions

6.1 (a) 1/4 (b) 1/2 (c) 1/2

6.2 (a) 26/52 · 25/51 (b) 26/52 · 25/51 + 26/52 · 25/51 (c) 26/51 (d) 25/51 (e) 26/51

6.3 0.7

6.4 0.2

6.5 1/3 (Note that 'exceed' means 'strictly greater than'.)

6.6 (a) $1/6 \cdot 1/6 = 6/36$ (b) No, because p(1st shows n ∧ sum is 7) = p(1st shows n) · p(sum is 7).

6.7 0.14

6.8 0.53

6.9 The test offered by the doctor is almost useless, because the probability is less than 1% that you suffer from the disease! (To be precise, it is 0.0018.)

6.10 0.47

6.11 By looking at Bayes' theorem you see that if $p(B) = 0$, then $p(B|A)$ is certain to be 0, because 0 divided by any positive number is 0.

6.12 If A is independent of A, then $p(A) = p(A \wedge A) = p(A) \cdot p(A)$. Hence, since the equation $x = x \cdot x$ has exactly two solutions in the interval between 0 and 1, namely 0 and 1, it follows that $p(A) = 0$ or $p(A) = 1$.

6.13 $p = \dfrac{1 - (365)!}{(365 - n)! \cdot 365^n}$ (Not all years have 365 days, but 2009 has.)

6.14 Oh, this is too difficult for me! Perhaps the answer is $1 - 1/81$?

7   The philosophy of probability

Philosophers disagree about what probability is. Somewhat roughly put, there are two camps in this debate, *objectivists* and *subjectivists*. Objectivists maintain that statements about probability refer to facts in the external world. If you claim that the probability is fifty per cent that the coin I hold in my hand will land heads up when tossed, then you are – according to the objectivist – referring to a property of the external world, such as the physical propensity of the coin to land heads up about every second time it is tossed. From an intuitive point of view, this seems to be a very plausible idea. However, subjectivists disagree with this picture. Subjectivists deny that statements about probability can be understood as claims about the external world. What is, for instance, the probability that your suitor will ask you to marry him? It seems rather pointless to talk about some very complex propensity of that person, or count the number of marriage proposals that other people make, because this does not tell us anything about the probability that *you* will be faced with a marriage proposal. If it is true that the probability is, say, 1/2 then it is true because of someone's mental state.

According to the subjective view, statements about probability refer to the degree to which the speaker believes something. When you say that the probability is 1/2 that your suitor will make a marriage proposal, this merely means that you believe to a certain degree that this proposal will be made, and the strength of your belief can be represented by the number 1/2; and if you were to start thinking that it was absolutely certain that you would get the offer, your subjective degree of belief would be 1. The scope of the subjective analysis can easily be extended to all kinds of examples, since virtually any probability statement can be interpreted as a claim about the speaker's subjective degree of belief. For instance, when you say that the probability is 1/2 that the coin will land heads up next time it is tossed, you are essentially referring to your subjective degree of belief that that event will occur.

Unfortunately, this way of distinguishing between objective and subjective views is just an approximation. Consider, for instance, the probability that I myself will believe something to a certain degree tomorrow. In this case, the objective interpretation would also refer to some fact about the speaker's mental state, rather than to facts about the external world. Another difficulty is that some probability statements seem to be neither objective nor subjective in the sense outlined here. Imagine, for instance, a scientist who claims that the probability that the law of nature N is true is 1/2 given the evidence currently available. Arguably, this is not a statement about a particular scientist's subjective degree of belief, nor about some random event in nature, because laws of nature are necessarily true. According to the *epistemic* interpretation of probability, some probabilities refer to the degree of support one statement gets from other statements.

In order to gain a better understanding of statements about probability the different interpretations have to be spelled out in more detail. In this chapter we outline three different versions of the objective interpretation (the *classical*, *frequentistic* and *propensity* interpretations), discuss the epistemic interpretation a bit more, and finally state and assess two versions of the subjective view.

7.1 The classical interpretation

The classical interpretation, advocated by Laplace, Pascal, Bernoulli and Leibniz, holds the probability of an event to be a fraction of the total number of possible ways in which the event can occur. For instance, as you roll a die exactly one of six sides will come up. The probability of getting an odd number is therefore 3/6, because out of the six possible outcomes exactly three are odd. Consider the following general definition.

(1) Probability = Number of favourable cases/Number of possible cases

It takes little reflection to realise that this interpretation presupposes that all possible outcomes are equally likely. This is by no means a tacit assumption. Advocates of the classical interpretation are very explicit about this point. According to these scholars, we determine the probability of an event by first reducing the random process to a set of equally likely outcomes; thereafter we count the number of outcomes in which the event occurs and divide by the total number of possible outcomes. Here is a passage from Laplace's *A Philosophical Essay on Probabilities.*

> The theory of chance consists in reducing all the events of the same kind to a certain number of cases equally possible, that is to say, to such as we may be equally undecided about in regard to their existence, and in determining the number of cases favourable to the event whose probability is sought. The ratio of this number to that of all the cases possible is the measure of this probability, which is thus simply a fraction whose numerator is the number of favourable cases and whose denominator is the number of all the cases possible. (Laplace 1814/1951: 7)

One may of course question whether it really makes sense to assume that it is always possible to reduce all events of the same kind into a set of 'equally possible' cases. Imagine that I roll a die I know is biased. Then it is no longer reasonable to maintain, to use Laplace's term, that it is 'equally possible' that each of the six sides will come up. Because if the die is biased it is of course 'more possible' that, say, a six will come up. However, contrary to what has been argued by some contemporary scholars, this does not necessarily show that there is anything wrong with the classical interpretation. Laplace and his allies could just point out that when confronted with a biased die the set of 'equally possible' cases is not identical with the six sides of the die. In that case, one has to reduce the set of cases further, until they become 'equally possible'. If you examine the die and discover that there is a piece of lead opposite the side marked six, making it twice as possible that you will roll a six as a one, you could individuate the set of possible cases in some other way. For instance, since it is *twice* as possible that you will roll a six as a one, it is *equally* possible that you will get a one in the first *or* in the second roll, as that you will roll a six in just one roll. Given that we have a method for deciding whether a pair of cases are equally possible, it seems that it will always be possible to divide the set of cases such that Laplace's criterion is met.

At this point one might ask what it *means* to say that two or more cases are 'equally possible'. To avoid ending up with a circular definition, the term 'equally possible' has to be defined without making any reference to the term 'probability'. This is where the real challenge begins. As long as we consider symmetric dice and well-balanced coins, our understanding of 'equally possible' is perhaps good enough. But what about the possibility that I might ask my partner for a divorce? Arguably, 'divorce' and 'no divorce' are not equally possible cases, but what does it *mean* to say that one of the cases is 'more possible' than the other?

A more technical problem with the classical interpretation is the following. If we suppose that the total number of cases is infinite, then the probability of every possibility will be zero, since the ratio between any finite number and infinity is zero. Imagine, for instance, that I ask you to pick a real number between 0 and 30; since there are uncountably many real numbers in every continuous interval the total number of cases will be infinite. In contemporary statistics examples of this sort are routinely analysed by introducing a so-called probability density function, but for Laplace it would make little sense to speak of probabilities if the set of possibilities is infinite.

Despite the objections raised against the classical interpretation, it has at least one merit: There is nothing wrong with it from a purely logical point of view, because it satisfies Kolmogorov's axioms. This is a necessary condition for any reasonable interpretation of the probability calculus. To show that the classical interpretation satisfies the axioms, we have to show that the axioms come out true when interpreted as stipulated by the interpretation. (This immediately implies that all theorems are true, since they were derived from the axioms.) To start with, recall the first axiom saying that all probabilities are numbers between 0 and 1, i.e. that $1 \geq p(A) \geq 0$. Since the set of favourable cases is a subset of the set of possible cases, it follows trivially that the ratio between the number of elements in the first set and the number of elements in the second set will always lie within the stipulated interval.

The second axiom holds that if a proposition is certain, then its probability is 1, and in the classical interpretation this means that all possible cases count as favourable cases; hence, the ratio between the two will be 1. Finally, according to the third axiom, $p(A \vee B) = p(A) + p(B)$ if A and B are mutually exclusive. In the classical interpretation all possible cases are (by definition) mutually exclusive, so it suffices to note that $\frac{i+j}{k} = \frac{i}{k} + \frac{j}{k}$ for all i, j and k.

7.2 The frequency interpretation

The frequency interpretation holds that the probability of an event is the ratio between the number of times the event has occurred divided by the total number of observed cases. Hence, if I toss a coin 1,000 times and it lands heads up 517 times then the relative frequency, and thus the probability, would be 517/1,000 = 0.517. Consider the following definition:

(2) Probability = Total number of positive instances/Total number of trials

To start with, let us verify that Kolmogorov's axioms come out as true when interpreted as stipulated by the frequency interpretation. The first axiom holds that all probabilities are numbers between 0 and 1. Since the total number of trials will always be at least as large as the total number of positive instances, the ratio between the two will always lie between 0 and 1. Furthermore, if the number of positive instances is equal to the total number of trials, then the ratio will equal 1, which shows that the second axiom comes out as true. Finally, the third axiom holds that $p(A \vee B) = p(A) + p(B)$ if A and B are mutually exclusive. The proviso that A and B are mutually exclusive means that no event belongs to A and B at the same time. Hence, the number of events that count as positive instances of $A \vee B$ equals the number of events that count as positive instances of A plus the number of events that count as positive instances of B. At the same time, the total number of trials is constant. Hence, since $\frac{i}{k} + \frac{j}{k} = \frac{i+j}{k}$ the third axiom also comes out as true.

In order to render the frequency interpretation more precise, a number of further issues must be clarified. To start with, suppose I toss the coin another 1,000 times, and that it lands heads up on 486 occasions. Does this imply that the probability has changed from 0.517 to 0.486? The physical constitution of the coin is the same, so anyone wishing to defend that view has quite a lot to explain.

But what if one claims that both series of tosses should be taken into account, and that the 'new' probability is therefore 517 + 486/2,000 ≈ 0.502? Now the same problem arises again, since it remains to be explained how the probability could change from 0.517 to 0.502 without there being any change at all in the physical constitution of the coin.

Note that in the frequency interpretation probability is always defined relative to some *reference class*. In the first example, it was tacitly assumed that the reference class is the set of 1,000 tosses, which was thereafter extended to a new reference class comprising 2,000 tosses. A major challenge for anyone seeking to defend the frequency interpretation is to specify which reference class is the relevant one and why. This problem becomes particularly pressing as the frequency interpretation is applied to unique events, i.e. events that only occur once, such as the US presidential election in 2000 in which George W. Bush won over Al Gore. The week before the election the probability was – according to many political commentators – about fifty per cent that Bush would win. Now, to which reference class does this event belong? If this was a *unique* event the reference class has, by definition, just one element, viz. the event itself. So if the US presidential election in 2000 was a unique event we are forced to the bizarre conclusion that the probability that Bush was going to win was 1/1, since Bush actually won the election (even though he did not get a majority of the votes and had to take his case to court) and the reference class comprises just a single element.

A possible reaction to this counter example is to argue that there was no 'real' probability to be determined in this case, precisely because the reference class is so small. But why would this be a good response? Exactly how large does a reference class have to be to count? Would a reference class of two, or three, or ten, elements be large enough? Between 1856 and 2009 there have been 38 presidential elections in the US, and republican candidates have won 18 of them. However, it would be strange to maintain that the probability that a republican will win next time is therefore 18/38. If someone like Forrest Gump were to run for presidency his probability of winning would certainly be much lower than 18/38. When determining the probability that the next president will be republican, the relative frequency of past events seems to be far less important than the political qualities of the present candidates.

The problem posed by unique events has inspired people to develop more sophisticated versions of the frequency interpretation. Venn, a pioneering nineteenth-century advocate of the frequency interpretation, argued that this interpretation makes sense only if the reference class is taken to be infinitely large. More precisely put, Venn argued that one should distinguish sharply between the underlying *limiting* frequency of an event and the frequency *observed* so far. It is not contradictory to suppose that the limiting frequency when tossing a coin is 1/2 even if the observed frequency is 0.517.

The limiting frequency is best thought of as the proportion of successful outcomes (proportion of heads, in the coin-tossing example) one *would* get if one were to repeat *one and the same* experiment infinitely many times. So even though the US presidential election in 2000 did *actually* take place just once, we can nevertheless *imagine* what would have happened had it been repeated infinitely many times. If the outcome had been the same every time, then we would then be justified in concluding that the election involved no genuine element of chance; otherwise we could simply determine the proportion of wins and losses.

That said, the limiting frequency view is usually thought to fit better with coin tossing and similar examples. We cannot actually toss a coin infinitely many times, but we could imagine doing so. This presupposes that the limiting frequency is thought of as an abstraction, rather as a series of events that takes place in the real world. Time has a beginning and end, so one and the same coin cannot be tossed infinitely many times, since it takes a couple of milliseconds to toss it.

The point that the limiting frequency is an abstraction rather than something that can be directly observed has some important philosophical implications. To start with, one can never be sure that a limiting relative frequency exists. When tossing a coin, the relative frequency of heads will perhaps never converge towards a specific number. In principle, it could oscillate for ever. A related problem is that the limiting relative frequency is inaccessible from an epistemic point of view, even in principle. Imagine that you observe that the relative frequency of a coin landing heads up is close to 1/2 in a series of ten million tosses. Now, this does not exclude that the true long-run frequency is much lower or higher than 1/2. No finite sequence of observations can prove that the limiting frequency is even close to the observed frequency. It is of course very unlikely that an average value calculated from ten million observations would differ dramatically from the true average value, but it is not impossible. Perhaps the coin would land tails almost every time if you were to toss it a billion times. A further problem is that, at least in principle, it is difficult to understand what it means to perform one and the same experiment more than once. Perhaps it makes most sense to say that all events are unique, even coin tosses. Although you may toss the same coin several times, the same particular toss never occurs twice. All coin tosses are unique.

7.3 The propensity interpretation

According to the propensity interpretation, probabilities can be identified with certain features of the external world, namely the propensity or disposition or tendency of an object to give rise to a certain effect. For instance, symmetrical coins typically have a propensity or disposition or tendency to land heads up about every second time they are tossed, which means that their probability of doing so is about one in two.

The propensity interpretation was famously developed by Karl Popper in the 1950s. His motivation for developing this view is that it avoids the problem of assigning probabilities to unique events faced by the frequency view. Even an event that cannot take place more than once can nevertheless have a certain propensity or tendency to occur. However, Popper's version of the theory also sought to connect propensities with long-run frequencies whenever the latter existed. Thus, his theory is perhaps best thought of as a hybrid between the two views. Contemporary philosophers have proposed 'pure' versions of the propensity interpretation, which make no reference whatsoever to long-run frequencies.

The propensity interpretation can be criticised for a number of reasons. First of all, the view needs to be more clearly stated. Exactly what is a propensity or disposition or tendency? And even if we know what it is, why should we believe that such entities exist? No one would be prepared to claim that a propensity can be directly observed, so this means that we have to rely on indirect evidence. However, for the reasons outlined above, no finite series of observations can ever prove beyond doubt that the real propensity of a coin landing heads up is even near the proportion of heads observed so far.

Perhaps the best response to the last objection is to distinguish sharply between the meaning of probability and the epistemic problem of how we ought to determine probabilities. Even if probabilities are in principle unknowable, it does not follow that they do not exist. Perhaps the real force of the objection is not its epistemic side, but rather the lack of semantic precision. It is simply not clear what it means to say that something is a propensity.

That said, the most well-known objection to the propensity interpretation is Humphreys' paradox. To state this paradox, recall that conditional probabilities can be 'inverted' by using Bayes' theorem, as explained in Chapter 6. Thus, if we know the probability of A given B we can calculate the probability of B given A, given that we know the priors. Now, the point made by Humphreys is that propensities cannot be inverted in this sense. Suppose, for example, that we know the probability that a train will arrive on time at its destination given that it departs on time. Then it makes sense to say that if the train departs on time, it has a propensity to arrive on time at its destination. However, even though it makes sense to speak of the inverted probability, i.e. the probability that the train departed on time given that it arrived on time, it makes no sense to speak of the corresponding inverted propensity. No one would admit that the on-time arrival of the train has a propensity to make it depart on time a few hours earlier.

The thrust of Humphreys' paradox is thus the following: Even though we may not know exactly what a propensity (or disposition or tendency) is, we do know that propensities have a temporal direction. If A has a propensity to give rise to B, then A cannot occur after B. In this respect, propensities function very much like causality; if A causes B, then A cannot occur after B. However, probabilities lack this temporal direction. What happens now can tell us something about the probability of past events, and reveal information about past causes and propensities, although the probability in itself is a non-temporal concept. Hence, it seems that it would be a mistake to identify probabilities with propensities.

7.4 Logical and epistemic interpretations

The logical interpretation of probability was developed by Keynes (1923) and Carnap (1950). Its basic idea is that probability is a logical relation between a hypothesis and the evidence supporting it. More precisely put, the probability relation is best thought of as a generalisation of the principles of deductive logic, from the deterministic case to the indeterministic one. For example, if an unhappy housewife claims that the probability that her marriage will end in a divorce is 0.9, this means that the evidence she has at hand (no romantic dinners, etc.) entails the hypothesis that the marriage will end in a divorce to a certain degree, which can be represented by the number 0.9. Coin tossing can be analysed along the same lines. The evidence one has about the shape of the coin and past outcomes entails the hypothesis that it will land heads up to a certain degree, and this degree is identical with the probability of the hypothesis being true.

Carnap's analysis of the logical interpretation is quite sophisticated, and cannot be easily summarised here. However, a general difficulty with logical interpretations is that they run a risk of being too dependent on evidence. Sometimes we wish to use probabilities for expressing mere guesses that have no correlation whatsoever to any evidence. For instance, I think the probability that it will rain in Johannesburg tomorrow is 0.3. This guess is not based on any meteorological evidence. I am just guessing – the set of premises leading up to the hypothesis that it will rain is empty; hence, there is no genuine 'entailment' going on here. So how can the hypothesis that it will rain in Johannesburg tomorrow be entailed to degree 0.3 or any other degree?

It could be replied that pure guesses are irrational, and that it is therefore not a serious problem if the logical interpretation cannot handle this example. However, it is not evident that this is a convincing reply. People do use probabilities for expressing pure guesses, and probability calculus can easily be applied for checking whether a set of such guesses is coherent or not. If one thinks that the probability for rain is 0.3 it would for instance be correct to conclude that the probability that there will be no rain is 0.7. This is no doubt a legitimate way of applying probability calculus. But if we accept the logical interpretation we cannot explain why this is so, since this interpretation defines probability as a *relation* between a (non-empty) set of evidential propositions and a hypothesis.

The debate on how to interpret the probability calculus can also be set up in a radically different way. According to this view the participants of the debate do not necessarily disagree about anything. This is because some statements about probability refer to objective (perhaps logical) properties, whereas other statements refer to subjective degrees of belief. It may very well hold true that some statements about coin tossing refer to objective features of the world, but this does not exclude the notion that other statements merely refer to the speaker's mental state. In this *ecumenical* approach, different versions of objective views can even coexist. In the modern literature, the logical interpretation is often called epistemic probability. When this label is used, it is generally assumed that probability statements tend to refer to different kinds of properties, and only probabilistic statements about the relation between a hypothesis and its evidence are called epistemic probabilities. Consider the following examples, and particularly compare the third with the first two.

(1) The probability that the coin will land heads up is 1/2. (Objective)
(2) The probability that Adam will divorce his wife Beatrice is 0.01. (Subjective)
(3) The probability that the law of nature N is true is 0.98 given present evidence. (Epistemic)

The third example is obviously a statement about how strong the relation is between a law of nature and some set of propositions supporting the law. Laws of nature are necessarily true or false, so it would make little sense to ascribe a probability to the law itself. Hence, philosophers such as Mellor and Williamson take this kind of example to show that some probabilities are epistemic, and that epistemic probabilities may coexist with other kinds of objective probability, as well as with subjective probability. Whether it is true that the probability is 0.98 that the law of nature holds true depends on whether the evidence actually supports the law to that degree, rather than on the constitution of the physical world or on the speaker's subjective degree of belief.

In the ecumenical approach, there is thus a genuine difference in meaning between these three examples. In the first example the speaker is talking about some objective feature of the world, and in the second she is talking about her subjective degree of belief, whereas the third example refers to some abstract epistemic relation between propositions.

Critics of the ecumenical approach could, of course, reply that *all* claims about probability are epistemic (or objective, or subjective). When talking about the probability of a coin landing heads up the speaker is merely talking about the extent to which the evidence about the coin supports the proposition that the coin will land heads up, and to say that my subjective probability that I will ask for divorce is 0.01 could be taken to mean that given my evidence, it is *rational* for me to deny that I will ask for divorce.

7.5 Subjective probability

The main idea in the subjective approach is that probability is a kind of mental phenomenon. Probabilities are not part of the external world; they are entities that human beings somehow create in their minds. If you claim that the probability for rain tomorrow is, say, 0.9 this merely means that your subjective degree of belief that it will rain tomorrow is strong and that the strength of this belief can be represented by the number 0.9. Of course, whether it *actually* will rain tomorrow depends on objective events taking place in the external world, rather than on your beliefs. So it is *probable* that it will rain tomorrow if and only if you believe to a certain degree that it will rain, irrespective of what actually happens tomorrow. However, this should not be taken to mean that any subjective degree of belief is a probability.

Advocates of the subjective approach stress that for a partial belief to qualify as a probability, one's degree of belief must be rational in a sense to be explained shortly.

Subjective probabilities can, of course, vary across people. Your degree of belief that it will rain tomorrow might be strong, at the same time as my degree of belief is much lower. This just means that our mental dispositions are different. The fact that we disagree does not in itself show that one of us must be wrong (even though the disagreement could only be a temporary one, given that both of us use Bayes' theorem for updating our beliefs, as explained in Chapter 6). When two decision makers hold different subjective probabilities, they just happen to believe something to different degrees. It does not follow that at least one person has to be wrong. Furthermore, if there were no humans around at all, i.e. if all believing entities were to be extinct, it would simply be false that some events happen with a certain probability, including quantum-mechanical events. According to the pioneering subjectivist Bruno de Finetti, "Probability does not exist."

Subjective views have been around for almost a century. De Finetti's pioneering work was published in 1931. Ramsey presented a similar subjective theory a few years earlier, in a paper written in 1926 and published posthumously in 1931. However, most modern accounts of subjective probability start off from Savage's theory (1954), which is more precise from a technical point of view. The key idea in all three accounts is to introduce an ingenious way in which subjective probabilities can be measured. The measurement process is based on the insight that the degree to which a decision maker believes something is closely linked to his or her behaviour. For example, if my belief is sufficiently strong that the food in the restaurant is poisoned I will not eat it, given that I desire to stay alive. Therefore, if it turns out that I do actually prefer to eat the food it seems that my subjective probability that there is no poison in it has to be fairly high.

The main innovations presented by Ramsey, de Finetti and Savage can be characterised as systematic procedures for linking probability calculus to claims about objectively observable behaviour, such as preference revealed in choice behaviour. Imagine, for instance, that we wish to measure Caroline's subjective probability that the coin she is holding in her hand will land heads up the next time it is tossed. First, we ask her which of the following very generous options she would prefer.

A "If the coin lands heads up you win a sports car; otherwise you win nothing."

B "If the coin *does not* land heads up you win a sports car; otherwise you win nothing."

Suppose Caroline prefers A to B. We can then safely conclude that she thinks it is *more probable* that the coin will land heads up rather than not. This follows from the assumption that Caroline prefers to win a sports car rather than nothing, and that her preference between uncertain prospects is entirely determined by her beliefs and desires with respect to her prospects of winning the sports car. If she on the other hand prefers B to A, she thinks it is *more probable* that the coin will not land heads up, for the same reason. Furthermore, if Caroline is indifferent between A and B, her subjective probability that the coin will land heads up is exactly 1/2. This is because no other probability would make both options come out as equally attractive, irrespective of how strongly she desires a sports car, and irrespective of which decision rule she uses for aggregating her desires and beliefs into preferences.

Next, we need to generalise the measurement procedure outlined above such that it allows us to always represent Caroline's degree of belief with precise numerical probabilities. To do this, we need to ask Caroline to state preferences over a *much larger* set of options and then *reason backwards*. Here is a rough sketch of the main idea: Suppose, for instance, that Caroline wishes to measure her subjective probability that her car worth $20,000 will be stolen within one year. If she considers $1,000 to be a fair price for insuring her car, that is, if that amount is the highest price she is prepared to pay for a gamble in which she gets $20,000 if the event S: "The car is stolen within a year" takes place, and nothing otherwise, then Caroline's subjective probability for S is $\frac{1,000}{20,000} = 0.05$, given that she forms her preferences in accordance with the principle of maximising expected monetary value. If Caroline is prepared to pay up to $2,000 for insuring her car, her subjective probability is $\frac{2,000}{20,000} = 0.1$, given that she forms her preferences in accordance with the principle of maximising expected monetary value.

Now, it seems that we have a general method for measuring Caroline's subjective probability: We just ask her how much she is prepared to pay for 'buying a contract' that will give her a fixed income if the event we wish to assign a subjective probability to takes place. The highest price she is prepared to pay is, by assumption, so high that she is indifferent between paying the price and not buying the contract.

The problem with this method is that very few people form their preferences in accordance with the principle of maximising expected monetary value. Most people have a decreasing marginal utility for money. However, since we do not know anything about Caroline's utility function for money we cannot replace the monetary outcomes in the examples with the corresponding utility numbers. (Note that we cannot apply von Neumann and Morgenstern's utility theory for determining Caroline's utility function, since it presupposes that the probability of each outcome is known by the decision maker.) Furthermore, it also makes little sense to *presuppose* that Caroline uses a specific decision rule, such as the expected utility principle, for forming preferences over uncertain prospects. Typically, we do not know anything about how people form their preferences.

Fortunately, there is a clever solution to all these problems. The basic idea is to impose a number of structural conditions on preferences over uncertain options. The structural conditions, or axioms, merely restrict what *combinations* of preferences it is legitimate to have. For example, if Caroline strictly prefers option A to option B in the sports car example, then she must not strictly prefer B to A. Then, the subjective probability function is established by reasoning backwards while taking the structural axioms into account: Since the decision maker preferred some uncertain options to others, and her preferences over uncertain options satisfy a number of structural axioms, the decision maker behaves *as if* she were forming her preferences over uncertain options by first assigning subjective probabilities and utilities to each option, and thereafter maximising expected utility. A peculiar feature of this approach is, thus, that probabilities (and utilities) are derived from 'within' the theory. The decision maker does not prefer an uncertain option to another *because* she judges the subjective probabilities (and utilities) of the outcomes to be more favourable than those of another. Instead, the well-organised structure of the decision maker's preferences over uncertain options logically implies that they can be described *as if* her choices were governed by a subjective probability function and a utility function, constructed such that a preferred option always has a higher expected utility than a non-preferred option. These probability and utility functions need not coincide with the ones outlined above in the car example; all we can be certain of is that there exist *some* functions that have the desired technical properties.

Box 7.1   Savage's representation theorem

Let us take a closer look at Savage's theory and discuss how it is related to von Neumann and Morgenstern's work. It will turn out that the main difference is the following: In von Neumann and Morgenstern's representation theorem it is assumed that the decision maker knows the probability of each outcome of the lotteries presented to her, whereas in Savage's theorem no such assumption about externally defined probabilities is required.

Let $S = \{s_1, s_2, ...\}$ be a set of states of the world with subsets $A, B, ...$ The latter can be thought of as events. If you roll two dice and get a double six, then this is an event having two states as elements. The set $X = \{x_1, x_2, ...\}$ is a set of outcomes. An example of an outcome is *I win $100*. Acts are conceived of as functions $f, g, ...$ from S to X. Put in words, this means that an act is a device that attaches an outcome to each state of the world. This is a very clever definition: An act is not a primitive concept; it is something that can be defined in terms of other concepts. Imagine, for instance, an act characterised by two states, *Heads* and *Tails*, and two outcomes, *Win $100* and *Lose $90*. If I am offered a bet in which I win $100 if the coin lands heads up and I lose $90 if it lands tails, then the act of taking on this bet is completely characterised by the following function f:

$f(Heads) = Win \$100$
$f(Tails) = Lose \$90$

Now, it seems reasonable to assume that a rational decision maker ought to be able to state pairwise preferences between this act and other acts, such as the act of taking on a bet in which one wins or loses twice as much. This new act, which we may call g, can be defined as follows.

$g(Heads) = Win \$200$
$g(Tails) = Lose \$180$

The expression $f \succeq g$ means that act f is *at least as preferred as* act g. (Indifference is the special case in which $f \succeq g$ and $g \succeq f$.) Now, it is helpful to introduce a couple of definitions. We say that f and g *agree* with each other in the set of states B if and only if $f(s) = g(s)$ for all $s \in B$. Put in words, this means that two acts agree in B if and only if they yield exactly the same outcome if event B occurs. Furthermore, $f \succeq g$ given B, iff B were known to obtain then f is weakly preferred to g; and B is *null*, iff $f \succeq g$ given

B for every f, g. It takes little reflection to realise that B is null in the case that B is considered to be virtually impossible to obtain.

Savage's first axiom is familiar from von Neumann and Morgenstern's work:

SAV 1 $\succeq$ *is a complete and transitive relation.*

The assumption that $\succeq$ is complete implies that it cannot be the case that the decision maker, after having considered two acts carefully, finds it impossible to tell which one she prefers the most. Personally, I know from introspection that this is descriptively false. I find some acts to be virtually impossible to compare. For example, even if I eventually manage to choose a wine from a wine list, this does not mean that my choice necessarily reveals any deeply considered preference. On many occasions, my 'choice' is merely the result of an unconscious random process; perhaps the probability was just 70% that I would choose the wine I actually did choose. However, according to Savage, his theory is normative and the fact that $\succeq$ is not always complete merely shows that I ought to *adjust* my preference according to his recommendation; otherwise I am not rational. The assumption that $\succeq$ is transitive may also be put in doubt, but let us for the time being assume that it is a reasonable normative requirement; anyone whose preferences are not transitive ought to *adjust* her preferences.

Here is Savage's second axiom, which is known as the sure-thing principle. It is less complex than the following formulation may indicate.

SAV 2   *If f, g and f', g' are such that*

1. *in $\neg B$, f agrees with g, and f' agrees with g',*
2. *in B, f agrees with f', and g agrees with g', and*
3. *$f \succeq g$;*

then $f' \succeq g'$.

The sure-thing principle is the most controversial of Savage's axioms. To understand why, we shall return to Allais' paradox, discussed in Chapter 4. Consider the decision problems in Table 7.1, in which exactly one winning lottery ticket will be drawn.

Suppose that you are first offered a choice between f and g, and thereafter another choice between f' and g'. Let B be the event of drawing

Table 7.1

	B (Ticket no. 1	Ticket no. 2–11)	¬B (Ticket no. 12–100)
Act f	$1M	$1M	$1M
Act g	$0	$5M	$1M
Act f'	$1M	$1M	$0
Act g'	$0	$5M	$0

a ticket numbered 1, 2, ..., or 11, and let ¬B be the event of drawing a ticket numbered 12, 13, ..., or 100. Note that in ¬B act f agrees with g, and f' agrees with g'. Furthermore, in B act f agrees with f', and g agrees with g'. When offered a choice between f and g many people feel that their preference is $f \succeq g$. Now, you can easily check that all the conditions of the sure-thing principle are satisfied, so it follows that $f' \succeq g'$. However, as explained in Chapter 4, many people would disagree with this conclusion. According to critics of the sure-thing principle, it is perfectly reasonable to accept that $f \succeq g$ but deny that $f' \succeq g'$. If you prefer f you will get a million for sure, but f' does not guarantee any money; hence, it might be worthwhile to take a risk and prefer g' over f'. However, let us accept the sure-thing principle for the sake of the argument, and see where Savage's axioms lead us.

Savage's third axiom seeks to establish a link between preferences over acts and preferences over outcomes. From a technical point of view, an outcome can be thought of as an act that is certain to yield a specific outcome. Imagine, for instance, that some act produces the same outcome no matter which state of the world is the true state. Call such acts *static*. Then, the third axiom stipulates that one static act is preferred to another if and only if the outcome produced by the first static act is preferred to the outcome produced by the second static act:

SAV 3   If $f(s) = x$, $f'(s) = x'$ for every $s \in B$, and B is not null, then $f \succeq f'$ given B, if and only if $x \succeq x'$.

Before we introduce the fourth axiom, yet another definition needs to be specified: B is *not more probable* than A if and only if f given A is at least as preferred as f given B (abbreviated as $f_A \succeq f_B$), provided that the outcome of f_A is preferred over the outcome of $f_{\neg A}$, $f_A = f_B$ and $f_{\neg A} = f_{\neg B}$. Note that this is a formalisation of the idea illustrated in the example with Caroline and the sports car. Now, it is natural to assume that:

SAV 4   *For every A and B it holds that A is not more probable than B or B is not more probable than A.*

The next axiom is a rather uncontroversial technical assumption. It states that there is some outcome that is strictly preferred to another (that is, it is not the case that every outcome is weakly preferred to every other outcome). This assumption was tacitly taken for granted in the sports car example, outlined above.

SAV 5   *It is false that for all outcomes $x, x', x \succeq x'$.*

It can be shown that a decision maker who accepts SAV 1–5 will be able to order all events A, B, ... from the least probable to the most probable. However, it does not follow that one is able to assign unique quantitative probabilities to these events. The next axiom, SAV 6, is adopted by Savage for establishing such a quantitative probability measure. Briefly put, the axiom says that every set of states can be partitioned into an arbitrary large number of equivalent subsets. This implies that one can always find a subset that corresponds exactly to each probability number.

SAV 6   *Suppose it is false that $f \succeq g$; then, for every x, there is a (finite) partition of S such that, if g' agrees with g and f' agrees with f except on an arbitrary element of the partition, g' and f' being equal to x there, then it will be false that $f' \succeq g$ or $f \succeq g'$.*

The six axioms stated above imply Theorem 7.1 below. The main difference compared to von Neumann and Morgenstern's theorem is that the probability function p is *generated from* the set of preferences over acts (uncertain prospects), rather than just taken for granted by the decision maker.

Theorem 7.1   (Savage's theorem) There exists a probability function p and a real-valued utility function u, such that:

(1) $f \succeq g$ if and only if $\int[u(f(s)) \cdot p(s)]ds > \int[u(g(s)) \cdot p(s)]ds$. Furthermore, for every other function u' satisfying (1), there are numbers $c > 0$ and d such that:

(2) $u' = c \cdot u + d$.

I will spare the reader the proof of Savage's theorem, which is rather complex. (An accessible version can be found in Kreps (1988).) The basic idea is similar to that used in the proof of von Neumann and Morgenstern's theorem. However, a few remarks on the interpretation of the theorem are appropriate.

Part (1) is the representation part. It guarantees the existence of a probability function representing degrees of belief (and a real-valued function representing utility), such that the agent can be described *as if* he was acting according to the principle of maximising expected utility. As explained above, no claim is made about what mental or other process *actually* triggered the decision maker's preferences. The theorem merely proves that a particular formal representation of those preferences is possible, which can be used for explaining and making predictions about future behaviour as long as all preferences remain unchanged. Part (2) is the uniqueness part, saying that utility is measured on an interval scale. It is exactly parallel to the uniqueness part of von Neumann and Morgenstern's theorem.

In the presentation outlined here, I have emphasised the similarities between Savage's and von Neumann and Morgenstern's theories. It is important to note that both theories use the same basic strategy for measuring a kind of mental state, even though Savage's slightly more complex theory manages to measure both utility and subjective probability at the same time – von Neumann and Morgenstern had nothing useful to say about where the probability function comes from, it is just taken for granted in their theorem.

7.5.1 State-dependent utilities

Imagine that you are standing next to James Bond. He is about to disarm a bomb, which has been programmed to go off within a few seconds. It is too late to escape; if Bond fails to disarm the bomb, both of you will die. Now ask yourself what your subjective probability is that Bond will manage to disarm the bomb before it goes off. Since you are now familiar with Savage's theory, you are prepared to state a preference between the following gambles:

A You win $100 if Bond manages to disarm the bomb and nothing otherwise.
B You win nothing if Bond manages to disarm the bomb and $100 if the bomb goes off.

Let us suppose that you prefer A over B. It follows directly from Savage's theory that your subjective probability that Bond will manage to disarm the bomb exceeds your subjective probability that he will fail to do so, for the same reason as in the sports car example. However, in the present example it seems reasonable to maintain that a rational agent will *always* prefer the gamble in which he wins $100 if the bomb is disarmed, no matter what he believes about the bomb. This is because if the bomb goes off money does not matter any more. (Once in Heaven, you will of course be served shaken martinis free of charge!) Hence, Savage's theory of subjective probability comes to the wrong conclusion. Even an agent who strongly believes that the bomb will go off will prefer the gamble in which he wins some money if the state he thinks is more unlikely occurs.

The problem illustrated by the James Bond example is that utilities are sometimes state-dependent, although Savage's theory presupposes that utilities are state-independent. That utilities are state-dependent means that the agent's desire for an outcome depends on which state of the world happens to be the true state.

A natural reaction to the James Bond problem is, of course, to argue that one should simply *add* the assumption that utilities have to be state-independent. Then the James Bond example could be ruled out as an illegitimate formal representation of the decision problem, since the utility of money seems to be state-dependent. However, the following example, originally proposed by Schervish, Kadane and Seidenfeld, shows that this is not a viable solution.

Suppose that the agent is indifferent between the three lotteries in Table 7.2. We then have to conclude that the agent considers the probability of each state to be 1/3. Also suppose that the agent is indifferent between the three lotteries in Table 7.3, in which the outcomes are given in Japanese yen. The three states are the same as before. Given that the decision maker's marginal utility for money exceeds zero, it follows that his subjective probability of s_1 is higher than his subjective probability of s_2, which is higher than his subjective probability of s_3. (Otherwise the expected utility

Table 7.2

	State 1	State 2	State 3
Lottery 1	$ 100	0	0
Lottery 2	0	$ 100	0
Lottery 3	0	0	$ 100

Table 7.3

	State 1	State 2	State 3
Lottery 1	¥100	0	0
Lottery 2	0	¥125	0
Lottery 3	0	0	¥150

of the three lotteries could not be equal.) It is therefore tempting to conclude that the decision maker has contradicted himself. It cannot be the case that the probability of each state is 1/3, at the same time as the probability of s_1 is higher than that of s_2.

However, suppose that the three states denote three possible exchange rates between dollars and yen. State s_1 is the state in which $100 = ¥100, s_2 is the state in which $100 = ¥125, and s_3 is the state in which $100 = ¥150. Obviously, this would render the decision maker's preferences perfectly coherent. By considering the hypothesis that the utility of money may be state-dependent for the agent, the hypothesis that the decision maker is not contradicting himself can be temporarily saved. However, important problems now arise. First of all, how could one tell from an *external* point of view whether utilities are state-dependent or not? If we have only observed the preferences stated in Table 7.2, what should we then say about the decision maker's subjective probability of the three states? And more importantly, is it *true* that s_1 is subjectively more probable that s_3 or not? Yes or no? As long as the probability function is not unique, this question will remain open.

It is beyond the scope of this textbook to review the vast literature on state-dependent utilities. Many articles have been published on this issue. Some authors have suggested axiomatisations that permit the derivation of a unique subjective probability function even in the case that the decision maker's utility function is state-dependent. This solution comes at a price, though. Such axiomatisations requires that the agent is able to state preferences over a much wider class of uncertain options than required by Savage.

7.5.2 The Dutch Book theorem

The Dutch Book theorem is commonly thought to provide an alternative route for justifying the subjective interpretation of probability. The theorem shows that subjective theories of probability are no less respectable from a mathematical point of view than objective ones.

A Dutch Book is a combination of bets that is certain to lead to a loss. Suppose, for instance, that you believe to degree 0.55 that at least one person from India will win a gold medal in the next Olympic Games (event G), and that your subjective degree of belief is 0.52 that no Indian will win a gold medal in the next Olympic Games (event $\neg G$). Also suppose that a cunning bookie offers you to bet on both these events. The bookie promises to pay you $1 for each event that actually takes place. Now, since your subjective degree of belief that G will occur is 0.55 it would be rational to pay up to $1 \cdot 0.55 = \$0.55$ for entering this bet. Furthermore, since your degree of belief in $\neg G$ is 0.52 you should be willing to pay up to $0.52 for entering the second bet, since $1 \cdot 0.52 = 0.52$. However, by now you have paid $1.07 for taking on two bets that are certain to give you a payoff of $1 *no matter what happens*. Irrespective of whether G or $\neg G$ takes place you will face a certain loss of $0.07. Certainly, this must be irrational. Furthermore, the *reason* why this is irrational is that your subjective degrees of belief violate the axioms of the probability calculus. As you should remember, we proved in Chapter 6 that the probability of $G \vee \neg G = 1$.

The Dutch Book theorem states that a decision maker's degrees of belief satisfy the probability axioms if and only if no Dutch Book can be made against her. This theorem, which was developed by Ramsey but independently discovered and proved more rigorously by de Finetti, is often taken to constitute an important alternative to Savage's axiomatic defence of subjective probability. The Dutch Book theorem emphasises the intimate link between preferences over uncertain options ('bets') and degrees of belief in a manner similar to Savage's axiomatic approach. However, in the Dutch Book theorem no utility function is derived; de Finetti simply took for granted that the decision maker's utility of money and other goods is linear. Many scholars have pointed out that this is a very strong assumption.

In order to get a better grasp of the Dutch Book theorem some qualifications have to be made. First, a *fair price* for a bet – say $100 for insuring a home worth $10,000 – is an amount such that the decision maker is equally willing to act as a player (the insured person) and as a bookie (the insurance company). Hence, it should not matter to the decision maker if she, depending on whether her home is destroyed or not, has to pay $100 and thereafter receives either $10,000 or nothing, or if she is the person who is being paid $100 and then has to pay either $10,000 or nothing. A further assumption is that there is exactly *one* fair price for every bet. If you, for instance, are equally willing to buy or sell a bet costing $100, your preference must be altered if the price is changed to $101. (To 'sell' a bet means that you take on a bet with a negative stake, such as −$100 instead of $100.)

The ratio between the fair price of a bet and the absolute value of the amount at stake is called the *betting quotient*. Hence, when you pay $100 for insuring a home worth $10,000 your betting quotient is $p = \dfrac{100}{10,000} = 0.01$. Now, to establish the Dutch Book theorem the decision maker has to announce his betting quotients for a fairly large number of bets: Let E be some initial set of statements about events x, x', ... that will or will not take place. Then, the decision maker has to announce his betting quotients for all elements in the set E^* comprising all standard truth-functional compounds of events in E, such as $\neg x$, $\neg x'$, $x \wedge x'$ and $x \vee x'$.

Next, the player is instructed to announce her betting quotients for all events E^*, that is, her subjective degrees of belief in all these events. At this point some important similarities with traditional betting stop: We assume that *the bookie is allowed to freely choose the stake S of each bet*, which may be either positive or negative. Hence, the player does not decide if she will, say, win $100 if x occurs or lose the same amount if this event takes place. We stipulate that the player must pay the bookie $p \cdot S$ for entering each bet, and that the player then receives S from the bookie if the event she is betting on occurs and nothing otherwise. The situation can be summarised in Table 7.4

The Dutch Book theorem consists of two parts. The first part shows that *if* the player's betting quotients violate the probability axioms, *then* she can be exploited in a Dutch Book. This means that it is pragmatically

Table 7.4

Player wins bet	Player loses bet
$S - p \cdot S$	$- p \cdot S$

irrational to violate the probability axioms. The second part of the
theorem shows that *no* Dutch book can be set up against a player whose
betting quotients satisfy the probability axioms; this indicates that anyone whose betting quotients satisfy the probability axioms cannot be
criticised for being irrational in a pragmatic sense. De Finetti only proved
the first part of the Dutch book theorem. The second part was proved
in the 1950s by other scholars. In Box 7.2 we recapitulate de Finetti's
part of the theorem.

The Dutch Book theorem is no doubt an important theorem. It shows
that there is an intimate link between rational degrees of beliefs and the
probability axioms. However, my personal view is that Savage's axioms
provide much stronger support for the subjective interpretation than
the Dutch Book theorem taken in isolation. One reason for this is that
the principle of maximising expected monetary value is simply taken for
granted in the setup of the theorem. It is not prima facie irrational to
suppose that some people are prepared to assign subjective degrees of belief
to events, and that these partial beliefs satisfy the probability axioms, even
though the agent refuses to evaluate bets by maximising expected value.
Perhaps the agent feels strongly committed to some other decision rule.
Another problem is, as highlighted by Savage, that it is rather implausible
to suppose that the player's utility for money is linear.

That said, one could of course *combine* the Dutch Book theorem with
Savage's axioms. Anyone who obeys Savage's axioms will act in accordance
with the expected utility principle. By replacing the monetary stakes with
utilities we avoid the assumption that the player's utility for money is
linear. However, the problem with this proposal is that we do not really
need the Dutch Book theorem if we have already accepted Savage's axiom: It
follows from Savage's axioms that the decision maker's partial beliefs
satisfy the probability axioms.

7.5.3 Minimal subjectivism

Both Savage's approach and the Dutch Book theorem seek to explicate
subjective interpretations of the probability axioms by making certain
claims about preferences over bets or other uncertain options. But, one
could ask, is really the relation between partial degrees of belief and
preferences that tight? A preference invariably requires a desire and a

Box 7.2   The Dutch Book theorem

Theorem 7.2   (de Finetti's part) If a player's betting quotients violate the probability axioms, then she can be exploited in a Dutch Book that leads to a sure loss.

Proof   We will consider each axiom in turn and show that if it is violated, then a Dutch Book can be made against the player.

The first axiom of the probability calculus holds that $1 \geq p(x) \geq 0$ for every event x. To start with, suppose that the player announces a betting quotient $p(x) > 0$. Then the bookie chooses some $S > 0$, and the player's net loss will be $S - p \cdot S < 0$ if x occurs, and $-p \cdot S < 0$ otherwise. Hence, player loses some money no matter whether x occurs or not. If $p(x) > 1$, then the bookie may choose $S = 1$, since, $1 - p \cdot 1 < 0$ and $-p \cdot 1 < 0$.

The second axiom holds that $p(x \vee \neg x) = 1$. (Strictly speaking, this is a theorem derived from the axiom.) Suppose that the player announces a betting quotient such that $p(x \vee \neg x) > 1$. Then the bookie chooses $S > 0$, and the player's net loss is $S - p \cdot S < 0$ since $x \vee \neg x$ is certain to occur, i.e. the player will 'win' the bet no matter what happens. If $p(x \vee \neg x) > 1$, then the bookie chooses $S < 0$, and the player's net loss is $S - p \cdot S < 0$. In both cases the player faces a certain loss.

The third axiom is the additivity axiom, according to which $p(x \vee x') = p(x) + p(x')$ given that x and x' are mutually exclusive. First, suppose that $p(x \vee x') > p(x) + p(x')$. Then, the bookie can simply choose $S = 1$ for the bet on $x \vee x'$, $S' = -1$ for the bet on x, and $S' = -1$ for the bet on x'. Then, since x and y are mutually exclusive events, only one of them can occur. For instance, if x but not x' occurs, then the player's payoff is $1 - p(x \vee x') \cdot 1$ from the bet on $x \vee x'$ and $-1 - p(x) \cdot -1$ from the bet on x'. Furthermore, he loses the bet on x', which incurs a loss of $-p(x') \cdot -1$. By adding these three terms we get: $1 - p(x \vee x') \cdot 1 - 1 - p(x) \cdot (-1) - p(x') \cdot (-1)$. This looks a bit messy, but this expression can be rewritten as $-p(x \vee x') + p(x) + p(x')$. However, the initial assumption that $p(x \vee x') > p(x) + p(x')$ entails that $-p(x \vee x') + p(x) + p(x')$ is negative, which means that the player will face a loss. The second case, $p(x \vee x') > p(x) + p(x')$, is analogous.   □

belief. When I tell the waiter that I would like tuna rather than salmon, then I believe that this will lead to an outcome (that I get some tuna) that I desire more than the other outcome (salmon). However, the converse does not hold. An agent can certainly believe something to a certain degree without having any preferences. The independence of beliefs and desires was stressed already by Hume, and few contemporary philosophers are willing to entirely reject his idea. Hence, it now seems natural to ask: Why on earth should a theory of subjective probability involve assumptions about preferences, given that preferences and beliefs are separate entities? Contrary to what is claimed by Ramsey, Savage and de Finetti, emotionally inert decision makers failing to muster any preference at all (i.e. people that have no desires) could certainly hold partial beliefs.

Here is a slightly different way of putting this point. If Ramsey, Savage and de Finetti are right, a decision maker cannot hold a partial belief unless he also has a number of desires, all of which are manifested as preferences over bets or uncertain options. However, contrary to what is claimed by Ramsey, Savage and de Finetti, emotionally inert agents could also hold partial beliefs. The assumption that there is a *necessary* link between probabilities and preferences (desires) is therefore dubious.

Does this argument pose any serious threat? The point is that the subjective theory of probability outlined above relies on an assumption that is not essential to the concept of probability, viz. the concept of desire. This shows that the *meaning* of the term 'subjective probability' is not captured by the traditional subjective analysis. Of course, it could be argued that in real life there are no emotionally inert agents, or at least no emotionally inert decision makers. However, this is hardly a relevant objection. For this argument to hold water, it is enough that the existence of an emotionally inert person is *possible*.

At this point Ramsey, Savage and de Finetti may object that they have never attempted to analyse the meaning of the term subjective probability. All they sought to do was to offer a procedure for explaining and predicting human behaviour. However, if this line of reasoning is accepted we must look elsewhere to find a definition of subjective probability. Some of Savage's contemporaries, e.g. Good, Koopman and DeGroot proposed *minimal* definitions of subjective probability that are not based on preferences over bets or uncertain options.

DeGroot's basic assumption is that decision makers can make *qualitative* comparisons between pairs of events, and judge which one they think is most likely to occur. For example, he assumes that one can judge whether it is *more*, *less* or *equally* likely, according to one's own beliefs, that it will rain today in Cambridge than in Cairo. DeGroot then shows that if the agent's qualitative judgements are sufficiently fine-grained and satisfy a number of structural axioms, then there exists a function p that (i) satisfies the probability axioms, and (ii) assigns real numbers between 0 and 1 to all events, such that one event is judged to be more likely than another if and only if it is assigned a higher number. So in DeGroot's minimal theory, the probability function is obtained by fine-tuning qualitative data, thereby making them quantitative. One may think of this as a kind of bootstrap approach to subjective probability. The probabilistic information was there already from the beginning, but after putting the qualitative information to work the information becomes quantitative.

Box 7.3   DeGroot's minimal subjectivism

In order to spell out DeGroot's minimal theory in more detail, let S be the sample and let E be a set of events to which probabilities are to be assigned, and let X, Y, ... be subsets of E. The relation 'more likely to occur than' is a binary relation between pairs of events; this relation is a primitive concept in DeGroot's theory. $X \succ Y$ means that X is judged to be more likely to occur than Y, and $X \sim Y$ means that neither $X \succ Y$ nor $Y \succ X$. The formula $X \succeq Y$ is an abbreviation for 'either $X \succ Y$ or $X \sim Y$, but not both'. Now consider the following axioms, which is supposed to hold for all X, Y, ... in E.

QP 1 $X \succeq \emptyset$ and $S \succ \emptyset$.

QP 1 articulates the trivial assumption that no event is less likely to occur than the empty set, and that the entire sample space is strictly more likely than the empty set.

QP 2   For any two events X and Y, exactly one of the following three relations hold: $X \succ Y$, or $Y \succ X$, or $X \sim Y$.

QP 2 requires that all events are comparable. This axiom resembles the completeness axiom in Savage's axiomatisation, according to which one must be able to rank any set of uncertain options without (explicitly) knowing the probabilities and utilities associated with their potential outcomes. However, QP 2 is less demanding than Savage's, since it does not involve any evaluative judgements.

QP 3   If X_1, X_2, Y_1 and Y_2 are four events such that $X_1 \cap X_2 = Y_1 \cap Y_2 = \emptyset$ and $Y_i \succeq X_i$ for $i = 1, 2$, then $Y_1 \cup Y_2 \succeq X_1 \cup X_2$. If, in addition, either $Y_1 \succ X_1$ or $Y_2 \succ X_2$, then $Y_1 \cup Y_2 \succ X_1 \cup X_2$.

In order to explain QP 3, suppose that some events can occur in either of two mutually exclusive ways, for example (1) 'the coin lands heads and you win a BMW' and (2) 'the coin lands tails and you win a BMW'; and (3) 'the coin lands heads and you win a Ford' and (4) 'the coin lands tails and you win a Ford'. Then, if (3) is more likely than (1) and (4) is more likely than (2), then it is more likely that you win a Ford than a BMW.

QP 4   If $X_1 \supset X_2 \supset \ldots$ and Y is some event such that $X_i \succeq Y$ for $i = 1, 2, \ldots$, then $X_1 \cap X_2 \cap \ldots \succeq Y$.

QP 4 guarantees that the probability distribution is countable additive. It follows from QP 4 that $Y \sim \emptyset$, because no matter how unlikely Y is ($Y \succ \emptyset$), it is impossible that for every n between one and infinity, the intersection of events $X_1 \cap X_2 \cap \ldots$ is at least as likely as Y, given that each X_n is more likely than X_{n+1}. For an intuitive interpretation, suppose that each X_n denotes an interval on the real line between n and infinity.

QP 5   There exists a (subjective) random variable which has a uniform distribution on the interval [0, 1].

QP 5 needs to be qualified. This axiom does not require that the random variable in question *really* exists. It is sufficient that the agent *believes* that this random variable exists. De Groot's approach to subjective probability theory makes no assumption about the nature of the external world – all that matters is the structure of internal subjective beliefs. QP 5 is thus consistent with the world being deterministic.

To understand what work is carried out by QP 5, suppose an agent wishes to determine her subjective probability for the two events 'rain here within the hour' and 'no rain here within the hour'. Then, since the set of events E only contains two elements, it is not possible to obtain a quantitative probability function by only comparing those two events. The set of events has to be extended in some way. QP 5 is the key to this extension. In a uniform probability distribution all elements (values) are equally likely. As an example, think of a roulette wheel in which the original numbers have been replaced with an infinite number of points in the interval [0, 1]. Then, by applying QP 5, the set of events can be extended to the union of the two original events and the infinite set of events 'the wheel stops at x ($0 \leq x \leq 1$)', etc.

Theorem 7.3 QP 1–5 are jointly sufficient and necessary for the existence of a unique function p that assigns a real number in the interval [0, 1] to all elements in E, such that $X \succeq Y$ if and only if $p(X) \geq p(Y)$. In addition, p satisfies Kolmogorov's axioms (see page 120).

Proof The proof consists of two parts. The first part shows that QP 1–5 imply the existence of a unique probability distribution. The second part verifies that the probability distribution satisfies Kolmogorov's axioms.

Part (1): The subjective probability function p is constructed by first applying QP 5, which entails that there exists a random variable which has a uniform distribution on the interval [0, 1]. Let $G[a, b]$ denote the event that the random variable is in the interval $[a, b]$. Then consider the following lemma:

Lemma   If x is any element in E, then there exists a unique number a^* ($1 \geq a^* \geq 0$) such that $x \sim G[0, a^*]$.

The proof of this lemma can be found in DeGroot's work. Now, if x is any element in A, we can apply the lemma and let $p(x)$ be defined as the number a^*. Hence, $x \sim [0, p(x)]$. It follows that $x \succeq y$ if and only if $p(x) \geq p(y)$, since $x \succeq y$ if and only if $G[0, p(x)] \succeq G[0, p(y)]$.

Part (2): It follows from the definition of p above that $p(x) \geq 0$ for every x. This verifies the first axiom of probability calculus. Moreover, $S = G[0, 1]$, where S is the entire sample space. This entails that $p(S) = 1$. This verifies the second axiom. To verify the third axiom, we have to show that $p\left(\bigcup_{i=1}^{n} x_i\right) = \sum_{i=1}^{n} p(x_i)$. To start with, consider the binary case with only two elements x_1 and x_2; that is, we want to show that if $x_1 \cup x_2 = \emptyset$, then $p(x_1 \cup x_2) = p(x_1) + p(x_2)$. In the first part of the proof we showed that $x_1 \sim G[0, p(x_1)]$. Hence, $x_1 \cup x_2 \sim G[0, p(x_1 \cup x_2)]$. According to a lemma proved by DeGroot, it also holds that $B \sim G[p(x_1), p(x_1 \cup x_2)]$. Now, note that $G[p(x_1), p(x_1 \cup x_2)] \sim G[0, p(x_1 \cup x_2) - p(x_1)]$. Also note that by definition, $B \sim [0, p(B)]$. Hence, $p(x_1 \cup x_2) - p(x_1) = p(x_2)$. By induction this result can be generalised to hold for any finite number of disjoint elements. □

Exercises

7.1 In Chapter 6 we showed that $p(A)+p(\neg A)=1$. Verify that this theorem comes out as true in the classical interpretation.

7.2 Summarise the major problems with the classical interpretation. In your view, which problem poses the greatest difficulty for the classical interpretation?

7.3 Prove that $p(A)+p(\neg A)=1$ in the frequency interpretation (without using Kolmogorov's axioms).

7.4 According to the frequency interpretation, probability is always defined relative to some reference class. What does this mean, and why might this spell trouble?

7.5 By definition, unique events (such as marriages or stock market collapses) never occur more than once. What do advocates of the following interpretations have to say about unique events: (a) the classical interpretation, (b) the frequency interpretation, (c) the propensity interpretation and (d) the subjective interpretation?

7.6 Consider the decision matrix below. You know that your marginal utility of money is decreasing, but you don't know by how much.

(a) Suppose that you are indifferent between Option 1 and Option 2. What can you conclude about your subjective probability for the two states?

(b) Suppose that you strictly prefer Option 1 to Option 2. What can you conclude about your subjective probability for State 1, as compared to State 2?

(c) Suppose that the $100 outcome is replaced with a black box with unknown contents, and that you feel indifferent between the two options. What can you conclude about your subjective probability for States 1 and 2?

	State 1	State 2
Option 1	$100	0
Option 2	0	$100

7.7 You are indifferent between receiving $10,000 for sure or five times that amount if and only if a Pakistani wins a gold medal in the next Olympic Games. Your utility for money is linear. What is your subjective probability that a Pakistani will win a gold medal in the next Olympic Games?

7.8 How did Savage define acts?

7.9 What is the sure-thing principle? Why is it controversial?

7.10 Explain briefly, using a non-technical vocabulary, how Savage manages to derive both a utility function and a subjective probability function from a set of preferences over uncertain prospects.

7.11 Explain the idea that the utility of an outcome may sometimes be state-dependent.

7.12 What is a Dutch Book?

7.13 Construct a Dutch Book to show that if A is impossible, then $p(A) = 0$.

7.14 Explain the difference between the betting approach to subjective probability and the qualitative approach taken by DeGroot. What are the merits and drawbacks of each approach?

Solutions

7.1 Let m be the number of cases that count as favourable with respect to A, and let n be the total number of cases. Then the number of favourable cases with respect to $\neg A$ is $n - m$. Hence, $\dfrac{m}{n} + \dfrac{n-m}{n} = \dfrac{n}{n} = 1.$

7.2 Explicit answers are outlined in the relevant section.

7.3 Let m be the number of positive instances of A, and let n be the total number of trials. Then the number of positive instances with respect to $\neg A$ is $n - m$. Hence, $\dfrac{m}{n} + \dfrac{n-m}{n} = \dfrac{n}{n} = 1.$

7.4 An answer is outlined in the relevant section.

7.5 Explicit answers are outlined in the relevant sections.

7.6 (a) $p(\text{State 1}) = 1/2$ and $p(\text{State 2}) = 1/2$ (b) You consider State 1 to be more probable that State 2. (c) It does not matter how much you like or dislike the prize in the black box, so $p(\text{State 1}) = 1/2$ and $p(\text{State 2}) = 1/2$.

7.7 1/5

7.8–7.12 Explicit answers are provided in the text.

7.13 Suppose that $p(A) > 0$. The bookie then chooses some $S > 0$. Then, since A is impossible the player will always lose any bet on A, so the bookie is certain to win S come what may.

7.14 An answer is outlined in the relevant section.

8   Why should we accept the preference axioms?

Here is a simple test of how carefully you have read the preceding chapters: Did you notice that some axioms have occurred more than once, in discussions of different issues? At least three preference axioms have been mentioned in several sections, viz. the transitivity, completeness and independence axioms. Arguably, this is all good news for decision theory. The fact that the same, or almost the same, axioms occur several times indicates that the basic principles of rationality are closely interconnected. However, this also raises a very fundamental concern: Do we really have any good reasons for accepting all these axioms in the first instance? Perhaps they are all false!?

In Table 8.1 we summarise the relevant axioms by using a slightly different notation than in previous chapters. Recall that the axioms are supposed to hold for all options x, y, z, and all probabilities p such that $1 > p > 0$.

A very simple standpoint – perhaps too simple – is that the preference axioms need no further justification, because we know the axioms to be true because we somehow grasp their truth immediately. However, many decision theorists deny they can immediately adjudicate whether a preference axiom is true, so it seems that some further justification is needed.

What more could one say in support of the axioms? According to an influential view, one has sufficient reason to accept a preference axiom if and only if it can somehow be *pragmatically justified*. Imagine, for instance, that you maintain that $x \succ y$ and $y \succ x$. Now suppose that you are holding x in your hand, and that I invite you to swap x for y on the condition that you pay me a small amount of money (one cent). Since you think y is better than x even after you have paid me a cent you agree to swap. I then invite you to swap back, that is, swap y for x, on the condition that you pay me one more cent. Since you *also* think that y is better than x, you agree to swap and pay me one cent. You are now back where you started, the only difference being

Table 8.1

Transitivity	If $x \succ y$ and $y \succ z$, then $x \succ z$
Completeness	$x \succ y$ or $y \succ x$ or $x \sim y$
Independence	If $x \succ y$, then $xpz \succ ypz$ (where xpz is a lottery that gives you x with probability p and z with probability $1 - p$.)

that you have two cents less. I now offer you to swap again. After a finite number of iterations you will be bankrupt, although you have not got anything in return. Hence, it is irrational to simultaneously prefer $x \succ y$ and $y \succ x$. This argument is called the *money-pump argument*, since it purports to show that an irrational decision maker can be 'pumped' of all her money.

The money-pump argument is a paradigmatic example of a pragmatic argument. Generally speaking, pragmatic arguments seek to show that decision makers who violate certain principles of rationality may face a decision problem in which it is *certain* that they will stand to loose, come what may. Thus, the decision maker will be worse off, as judged from her own perspective, no matter what happens. Next to the money-pump argument, the Dutch Book argument mentioned in Chapter 7 is perhaps the most influential pragmatic argument. However, note that pragmatic arguments need of course not be formulated in monetary terms. One could equally well construct a scenario in which the decision maker who follows her irrational preferences is certain to lose some ice cream, or happiness, or moral virtue, or whatever she cares about.

Pragmatic arguments provide the best support currently available for the axioms of decision theory. That said, it does not follow that this support is very strong. In this chapter we shall do two things in parallel. First, we shall take a closer look at pragmatic arguments for the transitivity, completeness and independence axioms. Second, while doing this we shall also discuss some alternative approaches to rational decision making. The presentation of alternative approaches will, however, be rather brief. The literature is still expanding, and many contributions require a lot of mathematical skills.

8.1 Must a rational preference be transitive?

Imagine that a friend offers to give you exactly one of her three love novels, x or y or z. You feel that you prefer x to y and y to z. However, does it follow from this that you must also prefer x to z? The transitivity axiom entails that the answer should be affirmative. Before we explain how it can be pragmatically justified, it is worth noting that it is sometimes considered to be a conceptual truth. Anyone who understands what it means to have a preference will eventually realise that preferences must not violate the transitivity axiom. To deny transitivity is like denying that a triangle must have three corners.

The downside of the conceptual argument is, of course, that it is sometimes very difficult to determine whether something is a conceptual truth. History is full of claims that were originally thought of as conceptual truths. For more than two thousand years Euclid's parallel axiom, saying that parallel lines will never cross, was thought to be a conceptual truth. However, the non-Euclidean geometry developed by the Hungarian mathematician Bolyai in the 1820s showed this to be false. It is *not* a conceptual truth that parallel lines will never cross.

The second argument for transitivity is based on a slightly more sophisticated version of the money-pump argument outlined above. This argument seeks to show that anyone whose preferences are cyclic (and hence not transitive) may end up in a situation in which it is certain that she will lose an infinite amount of money. The argument goes as follows. Imagine that your preference ordering over the three novels x, y and z is cyclic. You prefer x to y, and y to z, and z to x; i.e. $x \succ y \succ z \succ x$. Now suppose that you are in possession of z, and that you are invited to swap z for y. Since you prefer y to z, rationality obliges you to swap. So you swap, and temporarily get y. You are then invited to swap y for x, which you do, since you prefer x to y. Finally, you are offered to *pay a small amount*, say one cent, for swapping x for z. Since z is strictly better than x, even after you have paid the fee for swapping, rationality tells you that you should accept the offer. This means that you end up where you started, the only difference being that you now have one cent less. This procedure is thereafter iterated over and over again. After a billion cycles you have lost ten million dollars, for which you have got nothing in return.

Exactly what does the money-pump argument show? Obviously, the conclusion is *not* that a rational preference must be transitive. The conclusion is rather that if we permit cyclic preference orderings, then the contradiction outlined in Box 8.1 is unavoidable. Of course, one could deny transitivity without accepting cyclic preference orderings. Imagine, for

Box 8.1   The Money-pump argument

Does the money-pump argument provide a convincing justification of transitivity? In order to make a qualified judgement we first have to spell out the argument in more detail. Let $P_t(x)$ denote a predicate saying that it is permissible to choose x at time t. Now, the money-pump argument has two major premises. The first premise holds that value is choice-guiding:

Premise (i) In a pairwise choice between x and y at time t:

(i) If $x \succ y$, then $P_t(x) \wedge \neg P_t(y)$.
(ii) If $x \sim y$, then $P_t(x) \wedge P_t(y)$.

Note that clause (i) is very weak. If necessary, it could plausibly be strengthened to the claim that if $x \succ y$, then $O_t(x) \wedge \neg P_t(y)$, where $O_t(x)$ means that it is obligatory to choose x at t. Also note that we have not assumed that a rational preference must be complete. All we assume is that if x is preferred to y at t, then one is permitted to choose x and not y at t. (If the decision maker is indifferent between the two objects, she is of course permitted to choose any of them.)

The second premise of the money-pump argument is the principle of no payment. It holds that one is never rationally permitted to pay for x, if x could have been obtained for free. Of course, this premise is not universally valid, since things might change over time. Though you now could obtain a painting by the unknown artist Ossacip for free, it might be rational to pay a huge amount for it in the future, if Ossacip becomes as famous as Picasso. However, given that the principle of no payment is restricted to short periods of time, it seems hard to resist. In this context, the term 'short period' means that the utility of all objects must remain constant. Hence, the principle of no payment holds in the following situation: You will shortly receive exactly one object, x, or y, or z, and you know that these three are the only objects you can ever obtain. You are then offered a sequence of pairwise choices between the three objects. Now, if it is permitted to choose x at one point during the course of the experiment, then it can hardly be permitted to choose a strictly worse outcome, $x - \varepsilon$, at a later point.

To pay for an object x means that the agent gives up some small amount of utility ε in exchange for x. The formula $x - \varepsilon$ is a metaphorical expression of this exchange. Strictly speaking, $x - \varepsilon$ is an ordered pair $\langle x, -\varepsilon \rangle$, in which

x is the object received by the decision maker and $-\varepsilon$ is the amount the decision maker paid for getting x. Consider the following condition, which we assume holds for all x and t:

Premise (ii) If $P_t(x)$, then $\neg P_{t+n}(x-\varepsilon)$ for all $n \geq 1$.

In addition to the two premises stated above, we also need to make the following technical assumption: For every pair of distinct objects, if $x \succ y$ at t, then there exists some small amount of value ε such that $x - \varepsilon \succ y$ at t. Call this Assumption T.

We are now in a position to prove the following theorem:

Theorem 8.1 Premises 1 and 2 and Assumption T are logically inconsistent.

Proof Suppose for *reductio* that $x \succ y$ and $y \succ z$ and $x \succ z$. Let t_1, t_2, t_3 be three points in time and suppose that the agent is offered a choice between x at t_1, and between y and z at t_2, and between z and $x - \varepsilon$ at t_3. It follows from Premise (i) that $P_{t_1}(x)$ in the choice between x and y, since $x \succ y$. Now consider the choice made at t_3. Assumption T guarantees that ε can be chosen such that $x - \varepsilon \succ z$. Therefore, in a choice made at t_3 between z and $x - \varepsilon$, Premise (i) implies that $P_{t_3}(x - \varepsilon)$. However, since $P_{t_1}(x)$, as shown above, Premise (ii) implies that $\neg P_{t_3}(x - \varepsilon)$.   □

example, that you prefer x to y and y to z, and that you regard x and y to be incommensurable. (Let x be a large amount of money, and y health; money and health may be incommensurable.) This is a clear violation of transitivity, but in this case your preference ordering is not cyclic, and it is not certain that you can be money-pumped. Hence, it is difficult to see how the money-pump argument could support the transitivity axiom. It merely seems to exclude some, but not all, preference orderings that are incompatible with this axiom.

Moreover, it has been pointed out that a clever decision maker facing a money-pump would 'see what is in store for him' and reject the offer to swap and thus stop the pump. Thus, it cannot be taken for granted that someone who knows that he will be invited to swap several times should be prepared to do so, since he can simply foresee what is going to happen. In response to this objection, one could of course adjust the money-pump argument by stipulating that the agent has no reason to believe that he will be invited to make any more choices. Each new offer to swap comes as a surprise to him. All he knows is that x, y and z are the possible outcomes (plus a finite number of ε's). It still seems plausible to maintain that a reasonable theory of rational choice should guarantee that it is not permissible to choose $x - \varepsilon$ at t_3 if x could have been chosen at t_1. If one knew from the beginning that one could choose x and stay with it, and that there was no other strictly better object, why should one then accept to end up with $x-\varepsilon$? Several authors have also proposed other, more complicated ways of modifying the money pump, and thereby neutralise the objection that a clever decision maker would see what is in store for him.

8.2 Must a rational preference be complete?

The completeness axiom – holding that a rational decision maker must either prefer one object to another or be indifferent between the two – is quite strong. Consider, for instance, a choice between money and human welfare. Many authors have argued that it simply makes no sense to compare money with welfare. If the government could either save ten million dollars or let a citizen die, it is far from clear that a rational decision maker must be able to tell which option she prefers. Perhaps the cost of rescuing some fishermen from a sinking ship happens to be ten millions. Then, if you feel that you prefer saving a fellow citizen's life rather than spending ten million dollars less, consider a case in which the cost of saving the citizen is twenty, thirty or a hundred million dollars. If you accept the completeness axiom, you must maintain that there exists a precise cut-off point at which a certain amount of money becomes more preferable than saving a fellow citizen's life. (Unless one life saved outweighs every amount of money, which seems implausible.) This is because the completeness axiom entails that any pair of objects, no matter how disparate, can be compared. You must either prefer one to the other, or be indifferent between the two.

Can the completeness axiom be justified by some pragmatic argument? This depends on how one thinks incommensurability should be linked to rational choice. More precisely put, if one maintains that it is rationally permissible to swap between two incommensurable objects, then one can easily construct a money pump of the kind outlined above. Suppose that x and y are incommensurable, and that y^+ is strictly preferred to y. Now imagine a scenario in which you start with x, swap x for y, and then pay one cent for getting y^+. Finally, you swap y^+ for x. Now you are back where you started, the only difference being that you have one cent less. By iterating this procedure, a clever decision theorist can pump you of all your money. That said, if one *denies* that it is permissible to swap between incommensurable objects it would no longer be possible to construct this kind of money pump. There is currently no agreement on whether a decision maker who regards two objects as incommensurable should be permitted to swap. Therefore, it seems that the pragmatic justification of the completeness axiom is rather weak.

Having said all this, there is also an influential argument *against* the completeness axiom. This is the so-called small improvement argument. It runs as follows. Suppose that in a choice between ten million dollars and saving a human life, the decision maker neither (strictly) prefers ten millions to saving a human life, nor saving a human life to ten millions. Now, it would of course be a mistake to conclude that the decision maker must therefore be indifferent between ten millions and a human life. If one regards ten million dollars and a life saved to be equally valuable, it follows that if the ten million dollars or the life saved were modified by attaching an arbitrary small bonus, then the preference must tip over in favour of whichever object was thus modified. We stipulate that the addition of a small bonus, say one cent, to one of the alternatives would *not* have such a dramatic effect on the decision maker's preference. In a choice between a human life and ten million *plus* one cent, the decision maker would again be unable to muster a preference between the two alternatives. Hence, the decision maker was not indifferent between ten million dollars and a human life. So ten million dollars is neither better, nor worse, nor equally good as a human life. Hence, the two alternatives are incommensurable. (See Box 8.2.)

The small improvement argument is not uncontroversial. Some authors have questioned the premises it relies on, but it is beyond the scope of this short introduction to do justice to this debate. Suffice it to notice that a forceful objection to the small improvement argument can be derived from a view about preferences put forward by the economist Paul Samuelson in the 1930s. Inspired by the logical positivists, Samuelson argued that every statement about preferences must be revealed through choice behaviour. According to Samuelson's *revealed preference theory*, there is no conceptual

---

Box 8.2   The small improvement argument

The small improvement argument can easily be formalised. Consider three distinct objects, x, y and y^+, where y^+ is a small improvement of y (say, y plus one cent). Now, the point of the argument is to show that the four premises stated below together cast doubt on the completeness axiom. The first premise states that the decision maker neither strictly prefers ten millions to saving a human life, nor saving a human life to ten millions, while the second articulates the fact that ten millions plus one cent is preferred to ten millions. The third premise is a substantial logical assumption, while the fourth merely states that ten millions plus one cent is not preferred to saving a human life. (See Espinoza 2008.)

(1)	$\neg (x \succ y) \wedge \neg (y \succ x)$	premise
(2)	$y^+ \succ y$	premise
(3)	$(x \sim y) \wedge (y^+ \succ y) \rightarrow y^+ \succ x$	premise
(4)	$\neg (y^+ \succ x)$	premise
(5)	$\neg ((x \sim y) \wedge (y^+ \succ y))$	[from (3) and (4)]
(6)	$\neg (x \sim y)$	[from (2) and (5)]
(7)	$\neg (x \succ y) \wedge \neg (y \succ x) \wedge \neg (x \sim y)$	[from (1) and (6)]

The conclusion, (7), is an outright denial of the completeness axiom. Obviously, (7) follows from premises (1)–(4). Hence, the advocate of the completeness axiom must maintain that at least one of (1)–(4) is false.

---

space for incomplete preferences. Unless choices violate the *weak axiom of revealed preference* (WARP), they automatically reveal a preference. For example, if a decision maker is faced between two mutually exclusive and exhaustive alternatives x any y, and *chooses* x over y, then an external observer must conclude that x was at least as preferred as y. This means that it becomes very difficult to see what kind of behaviour could possibly correspond to incommensurability. If neither x nor y is chosen, e.g. if no choice at all is made, it is after all not a choice between two mutually exclusive and exhaustive options. This is why revealed preference theory does not acknowledge incommensurability as a possible preference relation. Whatever option the decision maker chooses, an advocate of revealed preference theory will argue that the chosen option was at least as preferred as the non-chosen one.

8.3 The multi-attribute approach

Faced with incomplete preference orderings, decision theorists sometimes propose a distinction between single- and multi-attribute approaches to decision theory. In a single-attribute approach, all outcomes are compared on a single utility scale. For example, in a decision between saving a group of fishermen from a sinking ship at a cost of ten million dollars or letting the fishermen die and save the money, the value of a human life will be directly compared with monetary outcomes on a single scale.

The multi-attribute approach seeks to avoid the criticism that money and human welfare are incommensurable by giving up the assumption that all outcomes have to be compared on a common scale. In a multi-attribute approach, each type of attribute is measured in the unit deemed to be most suitable for that attribute. Perhaps money is the right unit to use for measuring financial costs, whereas the number of lives saved is the right unit to use for measuring human welfare. The total value of an alternative is thereafter determined by aggregating the attributes, e.g. money and lives, into an overall ranking of the available alternatives.

Here is an example. Rachel has somehow divided the relevant objectives of her decision problem into a list of attributes. For illustrative purposes, we assume that the attributes are (i) the number of lives saved, (ii) the financial aspects of the decision, (iii) the political implications of the decision and (iv) the legal aspects of the decision. Now, to make a decision, Rachel has to gather information about the degree to which each attribute can be realised by each alternative. Consider Table 8.2, in which we list four attributes and three alternatives.

The numbers represent the degree to which each attribute is fulfilled by the corresponding alternative. For example, in the leftmost column the numbers show that the second alternative fulfils the first attribute to a higher degree than the first alternative, and so on. So far the ranking is

Table 8.2

	Attribute 1	Attribute 2	Attribute 3	Attribute 4
Alt. a_1	1	3	1	2
Alt. a_2	3	1	3	1
Alt. a_3	2	2	2	2

ordinal, so nothing follows about the 'distance' in value between the numbers. However, in many applications of the multi-attribute approach it is of course natural to assume that the numbers represent more than an ordinal ranking. The number of people saved from a sinking ship can, for instance, be measured on a ratio scale. This also holds true of the amount of money saved by not rescuing the fishermen. In this case, nothing prevents the advocate of the multi-attribute approach to use a ratio or interval scale if one so wishes.

Several criteria have been proposed for choosing among alternatives with multiple attributes. It is helpful to distinguish between two types of criteria, viz. additive and non-additive criteria. Additive criteria assign weights to each attribute, and rank alternatives according to the weighted sum calculated by multiplying the weight of each attribute with its value. The weights are real numbers between zero and one, which together sum up to one. Obviously, this type of criterion makes sense only if the degree to which each alternative satisfies any given attribute can be represented at least on an interval scale, i.e. if it makes sense to measure value in quantitative terms. Let us, for the sake of the argument, suppose that this is the case for the numbers in Table 8.2, and suppose that all attributes are assigned equal weights, i.e. $1/4$. This implies that the value of alternative a_1 is $1/4 \cdot 1 + 1/4 \cdot 3 + 1/4 \cdot 1 + 1/4 \cdot 2 = 7/4$. Analogous calculations show that the value of a_2 is 2, while that of a_3 is also 2. Since we defined the ranking by stipulating that a higher number is better than a lower, it follows that a_2 and a_3 are better than a_1.

Of course, one might question the method used for determining the weights. How can the decision maker rationally determine the relative importance of each attribute? An obvious answer is, of course, that one should ask the decision maker to *directly* assign weights to each attribute. A slightly more sophisticated approach is to let the agent make pairwise comparisons between all attributes, and thereafter normalise their relative importance. However, from a theoretical point of view the second approach is no better than the first. None of them can overcome the objection that such a direct approach makes the assignment of weights more or less arbitrary. It is indeed very optimistic to believe that decision makers can come up with adequate numerical weights. So perhaps the direct version of the multi-attribute approach is no better than the single-attribute approach: It is perhaps contentious to measure the utility of very different objects on a common scale, but it seems equally contentious to assign numerical weights to attributes as suggested here.

Most advocates of the multi-attribute approach favour an *implicit* strategy for making trade-offs between attributes. The basic idea has been imported from economic theory. Instead of directly asking the decision maker how important one attribute is in relation to another, we may instead ask him to state preferences among a set of hypothetical alternatives, some of which include more of one attribute but less of another. One can then indirectly establish indifference curves for all attributes, which show how much of one attribute the decision maker is willing to give up for getting one extra unit of another attribute. Given that the decision maker's preferences among alternatives are complete, asymmetric and transitive, it can then be proved that the decision maker behaves *as if* he is choosing among alternatives by assigning numerical utilities to alternatives, e.g. by assigning weights to the attributes and then adding all the weighted values.

That said, the implicit approach seems to put the cart before the horse in normative contexts. If one *merely* wishes to describe and predict the decision maker's future choices, it certainly makes sense to observe preferences among a set of alternatives and then assume that the trade-off rate between attributes implicit in those choices will be the same in future choices. However, if one wants to make a normative claim about what ought to be done, the implicit approach becomes more questionable. A decision maker who knows his preferences among alternatives with multiple attributes does not need any action guidance – so why divide the decision into a list of attributes in the first place? Furthermore, according to the implicit approach, it is false to say that one alternative act is better than another *because* its aggregated value is higher. All one is entitled to say is that the decision maker, as a matter of fact, behaves *as if* he was assigning numerical values to alternatives. From a normative point of view, this is of little help.

Non-additive aggregation criteria do not assume that the total value of an alternative can be calculated as a weighted sum. Consider, for example, the suggestion that the value of an alternative is obtained by multiplying the value of each attribute. Obviously, multiplicative criteria tend to put emphasis on the minimal degree to which each attribute is satisfied – a large number times zero is zero, no matter how large the large number is. Suppose, for example, that only two attributes are considered, and that the value of the first is zero and the value of the second is 999; then the product is of course also zero. An additive criterion would give at least some weight to the second value (999) so the overall value would in that case be strictly greater than zero, depending on the weight assigned to the second attribute.

Another example of a non-additive criterion is to impose *aspiration levels* for each attribute. The basic idea is that in case an attribute falls below a certain minimal level – the aspiration level – that alternative should be disregarded, no matter how good the alternative is with respect to other attributes. An obvious problem is, of course, that it might be difficult to specify the aspiration level. Furthermore, it might be questioned whether it is reasonable to assume that aspiration levels are sharp. Perhaps there are areas of vagueness, in which outcomes are neither below nor above the aspiration level.

Before closing this section, we shall consider a very general objection to multi-attribute approaches. According to this objection, there may exist several equally plausible but different ways of constructing the list of attributes. Sometimes the outcome of the decision process depends on which set of attributes is chosen – an alternative that is ranked as optimal according to a given decision criterion relative to one set of attributes might be ranked as suboptimal according to the same decision criterion relative to another equally reasonable set of attributes. Consider, for example, a medical products agency which is about to approve the new drug for the market. Imagine that the long-term effects of the drug are partly unknown, and that there is a concern that the drug may lead to increased cardiotoxicity. Cardiotoxicity might thus be taken to be a relevant attribute. However, let us suppose that for half of the population, the intake of the drug leads to an increased risk of cardiotoxicity, whereas for the other half it actually leads to a decreased risk. It is plausible to assume that the explanation of this difference is to be found in our genes. Now, if cardiotoxicity is conceived as one attribute, the drug might very well be approved by the regulatory agency, since no increased risk could be detected in a clinical trial – the two effects will cancel out each other. However, had we instead distinguished between two different attributes, 'cardiotoxicity for people with gene A' and 'cardiotoxicity for people with gene B' the result might have been quite different. If the increased risk for people with gene B is sufficiently large, the regulatory decision might be radically affected, depending on which criterion is used for aggregating risks. This example illustrates that advocates of multi-attribute approaches need a theory about how to individuate attributes, and if there is more than one way to individuate attributes they must not lead to conflicting recommendations. I leave it to the reader to decide whether the multi-attribute approach can be used to overcome the problem with incomplete preference orderings.

8.4 Must a rational preference satisfy the independence axiom?

We turn now to the independence axiom. Can this axiom be justified by a pragmatic argument? Many decision theorists have argued so. Their aim is to outline a scenario in which a decision maker who violates the independence axiom will stand to lose, come what may. In the standard version of the argument, the decision maker is asked to consider a sequential (dynamic) decision problem that is more complex than traditional money-pump arguments. As a point of departure we use Allais' famous decision problem discussed in Chapter 4. However, we shall present Allais' example in a slightly different way. Consider the following list of possible outcomes.

x	\$1,000,000
y	\$5,000,000 iff it rains tomorrow (R), and \$0 otherwise ($\neg R$)
z	\$0
$x + \varepsilon$	\$1,000,000 plus one cent
$z + \varepsilon$	\$0 plus one cent

We stipulate that the probability of rain tomorrow is 10/11, and that you know this. We also stipulate that there is an asymmetrical coin, which we know will land heads up (p) with probability 0.11. You are now presented with a set of choices over the outcomes listed above. First, you are asked to consider a choice between x and y. You choose x, i.e. $x \succ y$, since x gives you \$1M for sure. Next consider a lottery that gives you x with a probability of 0.11 and z otherwise. (We write xpz, as explained in Chapter 5.) How does xpz compare with ypz, i.e. a lottery that pays y with a probability of 0.11 and z otherwise? Clearly, ypz entitles you to a 11% chance of winning \$5M if it rains tomorrow, and nothing otherwise. Hence, the probability of winning \$5M is $0.11 \cdot 10/11 = 0.10$. Many people prefer a 10% chance of winning \$5M to a 11% chance of winning \$1M, and we assume that you are one of them. Hence, $ypz \succ xpz$. This means that your preferences are as follows.

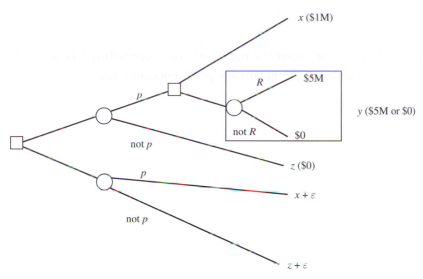

Figure 8.1   (After Rabinowicz 1995)

$$x \succ y, \text{ but } ypz \succ xpz \tag{1}$$

Clearly, you now violate the independence axiom, because independence implies that $xpz \succ ypz$ whenever $x \succ y$. Furthermore, given that the small bonus ε mentioned above is small enough, it will not reverse your preference between ypz and xpz. Hence, we may safely assume that:

$$ypz \succ xpz + \varepsilon \succ xpz \tag{2}$$

The preferences stated in (1) and (2) will lead to a certain loss. To see this, we now ask you to consider the sequential decision problem illustrated in Figure 8.1. Squares denote *choice nodes*, that is, points at which the decision maker makes her move (up or down). Circles stand for *chance nodes*, that is, points at which 'nature' makes its move (up or down). Note that the tree table is just a different way of illustrating Allais' paradox.

   A rational decision maker will solve this sequential decision problem by implementing a *plan*. In this context, a plan specifies exactly your moves at all choice nodes that you can reach with or without the help of the 'moves' made by nature at the chance nodes. Since you accept (1) and (2) you are committed to implement the following plan: You prefer $ypz \succ xpz + \varepsilon \succ xpz$, so you must therefore go *up* at the first (leftmost) choice node. Otherwise you will end up with $xpz + \varepsilon$. Given that p occurs and you reach the second (rightmost) choice node, the plan then specifies that you should go down, which will give you y. If p does not occur (that is, if the coin does not lands heads up), then you will get z. Let us call this plan Q. Plan Q is just a different way of expressing your preference $ypz \succ xpz + \varepsilon \succ xpz$. Given that ypz really is preferred to $xpz + \varepsilon$ and xpz, you are rationally committed to implement plan Q.

However, now suppose that you were to implement plan Q. Naturally, at the first choice node you would thus decide to go up. There are thus two cases to consider, depending on whether p occurs or not. Given that p occurs (the coin lands head up), you will reach the second choice node, at which you face a choice between x and y. Plan Q now dictates that you should go down, which will give you y. However, because of (1) we know that $x \succ y$, so *if you reach the second choice node you will no longer feel compelled to follow plan Q*, but instead go up and get x. Furthermore, if p does not occur (if the coin does not land heads up), you will get z. Compare all this with what would happen had you instead decided to go down in your first move. Clearly, had p occurred, you would have got $x + \varepsilon$ instead of x, and had p not occurred, then you would have got $z + \varepsilon$ instead of z. Hence, no matter whether p occurred or not, you would be better off by going *down* at the first choice node rather than up. This means that we have constructed an example in which a decision maker who acts in accordance with her own preferences will know for sure, before the first move is made, that the outcome will be suboptimal. Had you decided to go down at the first choice node, contrary to what is dictated by plan Q, you would have got whatever you would have got otherwise, *plus a small bonus ε*. Of course, if we repeat this thought experiment a large number of times, the loss can be made arbitrarily large.

Is this argument convincing? Many commentators have pointed out that a clever decision maker would foresee what is going to happen at the second choice node already at the first choice node, and thus go down instead of up at the first choice node to collect the bonus ε. Or, put in a slightly different way: A sophisticated decision maker would not face a sure loss of ε, because he would realise already from the beginning what was going to happen and take actions to avoid the trap. The notion of 'sophisticated' choice can be articulated in different ways. However, a popular strategy is to argue that a sophisticated decision maker should reason backwards, from the end nodes to the first choice nodes. This type of reasoning is known as backwards induction (cf. Chapters 1 and 11). Instead of solving the decision problem by starting at the beginning, one should instead start at all the end nodes and reason backwards. The argument goes as follows. Since the decision maker prefers x to y, he would go up at the second choice node. Hence, y can be ignored, since that outcome will never be chosen anyway. This means that the decision maker is essentially choosing between xpy and $(x + \varepsilon)p(z + \varepsilon)$ at the first choice node, and since he prefers $(x + \varepsilon)p(z + \varepsilon)$, we have identified a way of avoiding getting trapped. However, the backwards induction argument is controversial, for several reasons. First of all, backwards induction reasoning sometimes seems to lead to very odd conclusions, as illustrated in Chapter 11. It is not clear that this is a valid form of reasoning. Furthermore, an even more severe problem is that it is possible, as shown by several decision theorists, to construct sequential decision problems in which even a sophisticated decision maker, who reasons backwards from the end nodes to the first choice node, will be caught in a pragmatic trap. Such examples are quite complex, but the main idea is to show that even if one allows for backwards reasoning, a decision maker who violates the independence axiom may face situations in which it is certain that she will be worse off, by her own standards, when she acts on her preferences. That said, those alternative arguments have also been criticised, and there is currently no consensus on whether the independence axiom can be justified by some pragmatic argument.

8.5 Risk aversion

We shall finish this chapter with a discussion of risk aversion. Nearly all attempts to vindicate the independence axiom – and hence the expected utility principle – seek to appeal to intuitions about risk aversion. For instance, it was intuitions about risk aversion that motivated us to go up rather than down in the second choice node in the sequential choice describe above. Hence, a likely reason why one may prefer $1M to a lottery that yields either $5M or $0, even when the probability of winning $5M is high, is that a risk averter would feel that it is better to get a million for sure instead of facing a risk of getting nothing.

So what does it mean, more precisely, to say that a decision maker is risk averse? Unsurprisingly, several different notions of risk aversion have been proposed. The three most important notions can be summarised as follows.

1. Aversion against actuarial risks
2. Aversion against utility risks
3. Aversion against epistemic risks

Let us discuss each notion in turn. Kenneth Arrow, winner of the Nobel Prize in economics, characterised the first, actuarial notion of risk aversion like this: "[a] risk averter is defined as one who, starting from a position of certainty, is unwilling to take a bet which is actuarially fair". Thus, if you are offered a choice between three apples for sure, or a lottery that gives you a fifty-fifty chance of winning either six apples or nothing, you would prefer to get three apples for sure if you feel averse against actuarial risks. Note that the actuarial notion of risk aversion is *not* inconsistent with the expected utility principle (or the independence axiom). On the contrary, Arrow's basic idea is that intuitions about risk aversion can be accounted for *without* giving up the expected utility principle. By assuming that the decision maker's utility function for, say, apples is *concave* (i.e. slopes upwards but bends downwards) it follows that the expected utility of getting three apples for certain is higher than the expected utility of a fifty-fifty chance of winning either six apples or nothing. Take a look at Figure 8.2.

According to the graph in Figure 8.2, the utility of getting 3 apples is about 1.7 units of utility. However, the utility of winning 6 apples is about 2.4, so the expected utility of a lottery in which one has a fifty-fifty chance of winning either six apples or nothing is $2.4 \cdot 0.5 + 0 \cdot 0.5 = 1.2$, which is less than 1.7. Hence, a risk averse decision maker will prefer three apples to the lottery in which one wins either six apples or nothing. More generally

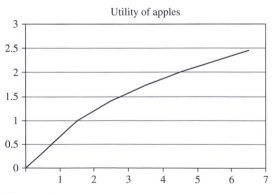

Figure 8.2

speaking, a decision maker whose utility function is concave (slopes upwards but bends downwards) will *always* prefer a smaller prize for certain to an actuarially fair lottery over larger and smaller prizes. You can see this by looking at the graph, while keeping in mind that the utility function has to be concave.

The actuarial notion of risk aversion can be mathematically defined as a claim about the shape of the decision maker's utility function, or, to be precise, as a claim about the ratio between second and first derivative of the utility function. (The value of the formula $-u''/u'$ serves a measure of actuarial risk aversion.) The 'flatter' the utility function is in a given interval, the more actuarially risk averse the decision maker is in that interval. It is beyond the scope of this introduction to explore the mathematical implications of this definition any further.

The notion of actuarial risk aversion has been widely employed in economic contexts. This is partly because it is consistent with the expected utility principle, but also because it yields predictions that can be easily tested. That said, scholars interested in normative aspects of decision theory complain that the actuarial notion cannot account for many central intuitions about risk aversion, such as the seemingly rational violation of the independence axiom explored by Allais. The idea underlying the second notion of risk aversion – aversion against utility risks – is that a risk averter ought to *substitute* the expected utility principle with some decision rule that is not only aversive to actuarial risks but also to (sufficiently large) utility risks. That is, instead of always maximising expected utility, one should sometimes (or always) apply some decision rule that puts more emphasis on avoiding bad outcomes. A potential loss of a large number of units of value cannot be compensated for by an equally large chance of winning the same number of units of value. A prominent example of a decision rule that is risk averse in this stronger sense is the maximin rule, discussed in Chapter 3. Other examples include rules that assign negative weight to the 'sufficiently bad' outcomes. Note, however, that such a weighting principle can be trivially transformed into a new utility scale, thereby making the new decision rule compatible with the expected utility principle.

This is not the right occasion for an in-depth discussion of the utility-based notion of risk aversion. However, note that if this notion is to deserve any attention it must fulfil at least two criteria. First, its decision rules must be genuinely different from the expected utility principle. More precisely put, one must show that the new rule cannot be reformulated as a version of the expected utility principle by e.g. transforming the utility scale. Second, advocates of the utility-based approach must also provide axiomatic justifications of their new decision rules, and this axiomatic support should be at least as robust as traditional axiomatisations of the expected utility principle. No decision theorists would seriously consider a decision rule that has not yet been axiomatised.

The third notion of risk aversion gets its force from intuitions about epistemic uncertainty. Imagine, for instance, that you are offered to bet on the outcome of three tennis matches, A, B and C. You know both players of match A very well, and you know for sure that they will play equally well – chance will decide who wins. You therefore assign probability 1/2 to each possible outcome of the match. However, match B is played between some players whom you know nothing about, although rumour has it that one player is much better than the other. The outcome will hardly be decided by chance, but you do not know which player is which, i.e. which is the better one. In an attempt to model this epistemic uncertainty, you decide to assign two parallel probability distributions to the outcome. According to the first distribution, the probability is 0.9 that player B_1 will win and 0.1 that B_2 will win; according to the second, the probability is 0.1 that player B_1 will win and 0.9 that B_2 will win. Finally, match C is also played between players whom you have never heard of before. In this case, you have no further information. As seen from your point of view, it simply makes no sense to assign any precise probabilities to the outcome of this match – all possible probability distributions are equally plausible.

At this point you are offered to bet on the outcome of each match. In each bet, you win 10 units of utility if the player you pick wins his match, and lose 9 units otherwise. Should you take on all three bets, or some of them, or none of them? It has been proposed by Gärdenfors and Sahlin (1988) that in a case like this it would be rational to apply *the maximin criterion for expected utilities (MMEU)*. According to this criterion, the alternative with the largest minimal expected utility ought to be chosen. Let us illustrate how this criterion works. In match B you operate with two parallel probability distributions. According to the first, the expected utility of bet B is $10 \cdot 0.9 + -9 \cdot 0.1 = 8.1$, but according to the second it is $10 \cdot 0.1 + -9 \cdot 0.9 = -7.1$. Hence, MMEU recommends you not to accept this bet. The minimal expected utility of refusing the bet is 0 (doing nothing), whereas the minimal expected utility of accepting the bet is −7.1.

Now consider match A. Here, you have only one probability distribution to consider, so the expected utility is $10 \cdot 0.5 + -9 \cdot 0.5 = 0.5$. Since $0.5 > 0$, you should accept the bet. Furthermore, in match C the MMEU rule collapses into the maximin rule. Since no probability distribution is more reasonable than another, the minimal expected utility is obtained when probability 1 is assigned to the worst outcome of the decision, which is exactly what is prescribed by the maximin rule. Hence, since the worst outcome of accepting the bet is that you lose 9 units, you should refuse the bet.

The MMEU rule is an attempt to capture intuitions about epistemic uncertainty. This rule is risk averse in the sense that it tells you to expect the worst. If the quality of the information at hand is low, this would indirectly justify a more cautious decision, and in the extreme case in which nothing is known about the probability of the outcomes one should reason as if the worst outcome was certain to occur. To some extent, Ellsberg's paradox discussed in Chapter 4 draws on the same intuition: Everything else being equal, one should prefer to take on bets in which one knows for sure what the probabilities are. However, one can of course question whether this intuition can really be rationally defended. Why should one expect the worst to happen? Why not the best? That is, in a state in which no probabilities are known, why accept the maximin rule rather than the maximax rule? The arguments against the maximin rule in Chapter 3 seem to be directly applicable also against the notion of epistemic risk aversion.

Exercises

8.1 Your preferences over a set of objects are not cyclic, i.e. it is *not* the case that there exists some x such that $x \succ \cdots \succ x$. Does it follow that your preferences satisfy the transitivity axiom?

8.2 (a) What is a pragmatic argument? (b) What is the money-pump argument?

8.3 The money-pump argument shows that an agent with cyclic preferences can be money-pumped. It does not entail that anyone who violates transitivity can be money-pumped. (a) Explain why! (b) Can the money-pump argument be strengthened, i.e. can this gap be filled? If so, how?

8.4 I prefer *x* to *y* and *y* to *z*, but I have no preference whatsoever between *x* and *z*. (I regard them to be incommensurable.) (a) Do my preferences violate transitivity? (b) Can I be money-pumped?

8.5 You will shortly be executed. However, before you are executed you will be given a free meal of your choice. The decision what to eat will be your very last one, and you know this for sure. Just as you are about to tell the prison guard what you would like to eat, you realise that if you were to act on your preferences you *could* be money-pumped, because your preferences are cyclic. However, for obvious reasons, you know for sure that you *will* not be money-pumped. Do you now have a reason to revise your preferences?

8.6 Why is Samuelson's theory of revealed preferences incompatible with the small improvement argument?

8.7 Consider the decision tree in Section 8.4 once again (Figure 8.1). Imagine that just before you make your choice at the first (leftmost) choice node, a being with perfect predictive powers offers to tell you (for a fee) whether it will rain tomorrow or not, i.e. whether *R* is true. Would you at this point, at the first choice node, be prepared to pay the being a fee for finding out the truth about *R*? If so, how much? (We assume that your utility of money is linear.)

8.8 You are offered a choice between (i) one left and a right shoe, and (ii) a fifty-fifty chance of getting nothing or ten left shoes. Since the utility of a left shoe in the absence of a right shoe is nil, you prefer (i) to (ii). Does this mean that you are risk averse in the actuarial sense, since you preferred two shoes to five expected shoes?

8.9 Karen's utility of money is $u = x^{1/2}$ and John's utility of money is $u = x^{1/3}$, where *x* is the current balance in each person's bank account. Who is most averse against actuarial risks, Karen or John, for large amounts of money?

Solutions

8.1 No.

8.2 (a) Pragmatic arguments aim to show that decision makers who violate certain principles of rationality can be presented with decision problems in which it is certain that they will stand to lose, come what may. (b) See Section 8.1.

8.3 (a) If you prefer x to y and y to z and regard x and y to be incommensurable, then you violate transitivity, but your preferences are not cyclic. (b) I leave it to the reader to make up his or her own mind about this.

8.4 (a) Yes (b) As the money-pump argument is traditionally set up, one cannot be money-pumped. However, this depends on what one thinks about rational decision making between incommensurable objects. If it is permissible to swap between incommensurable objects, and a mere possibility to end up slightly poorer than where one started is all that is required for the argument to be successful, then one can indeed be money-pumped.

8.5 I leave it to the reader to make up her own mind about this. (My personal view is: No!)

8.6 The theory of revealed preferences equates preference with choice. Thus, there is no room for a preference ordering being incomplete. One cannot choose one object over another, while claiming that the two objects were actually incommensurable.

8.7 If you find out that R is true you will certainly go up at the first choice node and down at the second, if you get there. If R is false, you will go down already at the first choice node. Hence, if R is true the expected utility of going up would be $0.11 \cdot 5M + 0.89 \cdot 0 = 0.55M$, and if R is false the expected utility of going down at the first choice node would be $0.11 \cdot (1M + \varepsilon) + 0.89(0 + \varepsilon) = 0.11M + \varepsilon$. You know that the probability that R is true is $10/11$. Hence, if you knew that you would be told about the truth value of R before you made your first choice, your payoff would be on average $10/11 \cdot 0.55M + 1/11 \cdot (0.11M + \varepsilon)$. However, as long as you do not know the truth of R before you make your choice at the first choice node, you will always go down, and can expect to win $0.11 \cdot (1M + \varepsilon) + 0.89 \cdot \varepsilon = 0.11M + \varepsilon$. The amount you should pay for finding out the truth about R is thus $10/11 \cdot 0.55M + 1/11 \cdot (0.11M + \varepsilon) - (0.11M + \varepsilon) = 10/11 \cdot 0.55M - 10/11 \cdot (0.11M + \varepsilon) = 10/11(0.44M + \varepsilon) = 0.4M + 10/11 \cdot \varepsilon$.

8.8 No. The relevant individuation of objects in this case is, arguably, to consider the number of pairs of shoes, rather than individual shoes. (That said, the question indicates that the actuarial notion of risk aversion is sensitive to deep metaphysical questions about the individuation of objects.)

8.9 Actuarial risk aversion is defined as $-u''/u'$. Hence, if $u = x^{1/2}$ then $u' = \frac{1}{2}x^{-1/2}$ and $u'' = -\frac{1}{4}x^{-3/2}$, so $-u''/u' = -\frac{1}{4}x^{-3/2} / \frac{1}{2}x^{-1/2} = -\frac{1}{2}x^{-3}$. Furthermore, if $u = x^{1/3}$ then $u' = \frac{1}{3}x^{-2/3}$ and $u'' = -\frac{2}{9}x^{-5/3}$, so $-u''/u' = -\frac{2}{9}x^{-5/3} / \frac{1}{3}x^{-2/3} = -\frac{2}{3}x^{-10/3}$. It follows that John is more risk averse than Karen for all $x > 1$.

9 Causal vs. evidential decision theory

The focus of this chapter is on the role of *causal processes* in decision making. In some decision problems, beliefs about causal processes play a significant role for what we intuitively think it is rational to do. However, it has turned out to be very hard to give a convincing account of what role beliefs about causal process should be allowed to play. Much of the discussion has focused on a famous example known as Newcomb's problem. We shall begin by taking a look at this surprisingly deep problem.

9.1 Newcomb's problem

Imagine a being who is very good at predicting other people's choices. Ninety-nine per cent of all predictions made by the being so far have been correct. You are offered a choice between two boxes, B_1 and B_2. Box B_1 contains $1,000 and you know this, because it is transparent and you can actually see the money inside. Box B_2 contains either a million dollars or nothing. This box is not transparent, so you cannot see its content. You are now invited to make a choice between the following pair of alternatives: You either take what is in both boxes, or take only what is in the second box. You are told that the predictor will put $1M in box B_2 *if and only if* she predicts that you will take just box B_2, and nothing in it otherwise. The predictor knows that you know this. Thus, in summary, the situation is as follows. First the being makes her prediction, then she puts either $1M or nothing in the second box, according to her prediction, and then you make your choice. What should you do?

Alternative 1   Take box B_1 ($1,000) and box B_2 (either $0 or $1M).
Alternative 2   Take only box B_2 (either $0 or $1M).

This decision problem was first proposed by the physicist William Newcomb in the 1960s. It has become known as Newcomb's problem, or the predictor's paradox. Philosopher Robert Nozick was the first author to discuss it in print. (According to legend, he learnt about the problem from a mutual friend of his and Newcomb's at a dinner party.) Nozick identified two different, but contradictory, ways of reasoning about the problem. First, one can argue that it is rational to take both boxes, because then you get the $1,000 in the first box, and whatever amount of money there is in the second. Since the $1M either is or is not in the second box as you make your choice, the fact that the predictor has made a prediction does not make any difference. The predictor makes her prediction, and adjusts the amounts of money in the second box accordingly, before you make your choice. She certainly does not have any magical powers, so she cannot adjust the amount in the second box *now*, as you are about to choose. The $1M either is in the box or is not, no matter what you decide to do. Therefore you should better take both boxes, because then you will be better off, come what may. This line of reasoning can be seen as a straight-forward application of the dominance principle: Taking two boxes dominates taking just one.

However, if you grab both boxes you haven't yet really taken all relevant information into account. It seems rational to also consider the fact that the predictor has predicted your choice, and adjusted the amounts of money in the second box accordingly. As explained above, the probability that the predictor has made a correct prediction is 99%. Hence, if you take both boxes she has almost certainly predicted this, and hence put $0 in the second box. However, if you take only the second box, the pre-dictor has almost certainly predicted that decision correctly, and conse-quently put $1M in the second box. Hence, by taking just the second box it is much more likely that you will become a millionaire. So why not take only the second box? This line of reasoning can be seen as a straightforward application of the principle of maximising expected utility. Consider the decision matrix in Table 9.1.

Table 9.1

	Second box contains $1M	Second box is empty
Take second box only	$1M (prob. 0.99)	$0 (prob. 0.01)
Take both boxes	$1M + $1,000 (prob. 0.01)	$1,000 (prob. 0.99)

To keep things simple, we assume that your utility of money is linear, i.e. we assume that dollars can be directly translated into utilities. (This is a non-essential assumption, which could easily be dropped at the expense of some technical inconvenience.) Hence, the expected utility of taking only the second box is as follows:

Take box 2: $0.99 \cdot u(\$1M) + 0.01 \cdot u(\$0) = 0.99 \cdot 1,000,000 + 0.01 \cdot 0 = 990,000.$

However, the expected utility of taking both boxes is much lower:

Take box 1 and 2: $0.01 \cdot u(\$1M) + 0.99 \cdot u(\$0) = 0.01 \cdot 1,000,000 + 0.99 \cdot 0$
$$= 10,000.$$

Clearly, since $990,000 > 10,000$, the principle of maximising expected utility tells you that it is rational to only take the second box. Hence, two of the most fundamental principles of decision theory – the dominance principle and the principle of maximising expected utility – yield conflicting recommendations. So what should a rational decision maker do?

When presented with Newcomb's problem, students (and other normal people) often come up with a wide range of suggestions as to what may have gone wrong. Some of those reactions indicate that the commentator has not correctly understood how the problem has been set up. Therefore, to steer clear of a number of trivial mistakes, consider the suggestion that one cannot know *for sure* that the probability is 0.99 that the predictor has made an accurate prediction since such good predictors do not exist in the real world. At a first glance, this may appear to be a reasonable reaction. However, anyone who thinks a bit further will sees that this point is irrelevant. Decision theorists seek to establish the most general principles of rationality, and we can surely *imagine* a being with the very impressive predictive powers stipulated in the example. For instance, suppose that a thousand people have been offered the same choice before. Nearly all who choose one box are now millionaires, whereas almost everyone who took both boxes got just $1,000. Isn't this a good reason for trusting the predictive powers of the being?

Another common source of misunderstanding is that Newcomb's problem somehow presupposes that the predictor has magical powers, or can alter the past. This is simply not the case, however. The predictor is just a very good predictor. In all other respects she is much like you and me. Here is an analogous example, proposed by James Joyce, of a situation that is similar to Newcomb's problem, but which is less odd from a metaphysical point of view. Imagine that a team of researchers has recently discovered that students who read *An Introduction to Decision Theory* by Martin Peterson always have a firm interest in decision theory. (Indeed, anyone who reads such a terribly boring book must have a truly outstanding interest in the subject!) Unfortunately, the researchers have also discovered that nearly all students of decision theory end up leading miserable lives, since they spend too much time thinking about abstract decision rules and fail to appreciate the less theoretical aspects of human flourishing. More precisely put, the researchers have discovered that students who read Section 9.2 of *An Introduction to Decision Theory* always pass the exam at the end of the course, whereas those who stop reading at Section 9.1 always fail. Furthermore, it is also known that 99% of all students taking courses in decision theory have a gene that causes two separate effects. First, the gene causes the desire to read Section 9.2 of *An Introduction to Decision Theory*. Second, the gene also causes their miserable lives. Thus, given that *you* have the gene, if you read Section 9.2 you know that you will lead a miserable life, although you will of course pass the exam. If you do not have the gene and continue to read you will lead a normal life and pass the exam, but if you stop reading now you will fail the exam. Naturally, you prefer to pass the exam rather than to fail, but you much prefer to lead a happy life not passing the exam instead of passing the exam and leading a miserable life. Is it rational for you to read Section 9.2? You are now at Section 9.1. Think carefully before you decide.

9.2 Causal decision theory

So you decided to read this section!? People advocating *causal decision theory* will congratulate you. Briefly put, causal decision theory is the view that a rational decision maker should keep all her beliefs about causal processes fixed in the decision-making process, and always choose an alternative that is optimal according to these beliefs. Hence, since you either have the gene or not it is better to read this section, because then you will at least pass the exam. If you have the gene, it is not you reading this section that causes your miserable life. It is the gene that causes this unfortunate outcome – and there is nothing you can do to prevent it. Furthermore, if you belong to the 1% minority who end up leading normal lives it is of course better to pass the exam than to fail, so in that case it is also rational to read this

Table 9.2

	Gene	No gene
Read Section 9.2	Pass exam & miserable life	Pass exam & normal life
Stop at Section 9.1	Fail exam & miserable life	Fail exam & normal life

section. Hence, it is rational to read this section no matter whether you have the gene or not. The decision matrix in Table 9.2 illustrates how we established this conclusion.

According to causal decision theory, the probability that you have the gene given that you read Section 9.2 is equal to the probability that you have the gene given that you stop at Section 9.1. (That is, the probability is independent of your decision to read this section.) Hence, it would be a mistake to think that your chances of leading a normal life would have been any higher had you stopped reading at Section 9.1. The same line of reasoning can be applied to Newcomb's problem. Naturally, causal decision theory recommends you to take two boxes. Since you know that the predictor has already placed or not placed $1M in the second box, it would be a mistake to think that the decision you make *now* could somehow affect the probability of finding $1M in the second box. The causal structure of the world is forward-looking, and completely insensitive to past events.

Causal decision theory leads to reasonable conclusions in many similar examples. For instance, imagine that there is some genetic defect that is known to cause both lung cancer and the drive to smoke, contrary to what most scientists currently believe. However, if this new piece of knowledge were to be added to a smoker's body of beliefs, then the belief that a very high proportion of all smokers suffer from lung cancer should not prevent a causal decision theorist from starting to smoke, because: (i) one either has that genetic defect or not, and (ii) there is a small enjoyment associated with smoking, and (iii) the probability of lung cancer is not affected by one's choice. Of course, this conclusion depends heavily on a somewhat odd assumption about the causal structure of the world, but there seems to be nothing wrong with the underlying logic. Given that lung cancer and the drive to smoke really have the same cause, then the action prescribed by the causal decision theorist seems to be rational.

Having said all this, it is helpful to formulate causal decision theory in more detail in order to gain a better understanding of it. What does it mean to say, exactly, that rational decision makers should do whatever is most likely to bring about the best expected result, while *holding fixed all views about the likely causal structure of the world*? The slogan can be formalised as follows. Let $X \mathbin{\square\!\!\rightarrow} Y$ abbreviate the proposition 'If the decision maker were to do X, then Y would be the case', and let $p(X \mathbin{\square\!\!\rightarrow} Y)$ denote the probability of $X \mathbin{\square\!\!\rightarrow} Y$ being true. In the smoking case, the causal decision theorist's point is that $p(\text{Smoke} \mathbin{\square\!\!\rightarrow} \text{Cancer})$ is equal to $p(\neg\text{Smoke} \mathbin{\square\!\!\rightarrow} \text{Cancer})$. Hence, it is better to smoke, since one will thereby get a small extra bonus. There is of course a strong statistical correlation between smoking and cancer, but according to the assumption of the example it is simply false that smoking causes cancer. Therefore it is better to smoke, since that is certain to yield a small extra enjoyment. Furthermore, in the exam case the causal decision theorist will point out that $p(\text{Read Section 9.2} \mathbin{\square\!\!\rightarrow} \text{Miserable life})$ is equal to $p(\text{Stop at Section 9.1} \mathbin{\square\!\!\rightarrow} \text{Miserable life})$. You either have the gene or you don't. Hence, the decision to continue reading was rational since it entails that you will at least pass the exam. Finally, in Newcomb's problem, $p(\text{Take both boxes} \mathbin{\square\!\!\rightarrow} \$1M)$ is equal to $p(\text{Take second box only} \mathbin{\square\!\!\rightarrow} \$1M)$, and since it is certain that you will get an extra \$1,000 if you take both boxes you should do so.

9.3 Evidential decision theory

The causal analysis of Newcomb-style problems has been influential in academic circles in recent years, although it is by no means uncontroversial. Some decision theorists claim there are cases in which causal decision theory yields counterintuitive recommendations. Imagine, for instance, that Paul is told that the number of psychopaths in the world is fairly low. The following scenario would then cast doubt on the causal analysis.

> Paul is debating whether to press the 'kill all psychopaths' button. It would, he thinks, be much better to live in a world with no psychopaths. Unfortunately, Paul is quite confident that only a psychopath would press such a button. Paul very strongly prefers living in a world *with* psychopaths to dying. Should Paul press the button? (Egan 2007: 97)

Intuitively, it seems reasonable to maintain that Paul should not press the button. If he presses, it is probable that he will die, since nearly everyone who presses is a psychopath. However, causal decision theory incorrectly implies that Paul should press the button. To see why, note that p(press button $\square\rightarrow$ dead) is much lower than p(press button $\square\rightarrow$ live in a world without psychopaths). This is because Paul either is or is not a psychopath, and the probability of the two possibilities does not depend on what he decides to do. In order to support this claim further, it may be helpful to dress up the example with some numbers. Let us, somewhat arbitrarily, suppose that the probability of p(press button $\square\rightarrow$ dead) is 0.001, since there are very few psychopaths, whereas p(press button $\square\rightarrow$ live in a world without psychopaths) is 0.999. Let us furthermore suppose that the utility of death is −100, and that the utility of living in a world without psychopaths is +1, whereas the utility of not pressing the button and living in a world with psychopaths is 0. According to the causal decision theorist, the expected utilities of the two alternatives can thus be calculated as follows.

Press button: p(press button $\square\rightarrow$ dead) $\cdot$ u(dead) $+$ p(press button $\square\rightarrow$ live in a world without psychopaths) $\cdot$ u (live in a world without psychopaths)$= (0.001 \cdot -100) + (0.999 \cdot 1) = 0.899$

Do not press button: p(do not press button $\square\rightarrow$ live in a world with psychopaths) $\cdot u$ (live in a world with psychopaths) $= 1 \cdot 0 = 0$

If expected utilities are calculated in this way, it follows that pressing the button is more rational than not pressing. However, as argued above, it is not clear that this is the right conclusion. We intuitively feel that it would be better not to press, even after a quick rehearsal of the argument proposed by the causal decision theorist. Some decision theorists, known as *evidential* decision theorists, explicitly deny the causal analysis. Evidential decision theorists claim that it would be rational not to press. The gist of their argument is that they think causal decision theorists calculate probabilities in the wrong way. In the psychopath case, it is almost certain that if you press the button then you are a psychopath yourself, and this insight must somehow be accounted for. Evidential decision theorists agree with causal decision theorists that pressing the button does not *cause* any psychiatric disease, but if you were to press the button you would indirectly learn something about yourself that you did not already know, namely that you are a psychopath. This piece of extra information cannot be accounted for in any reasonable way by the causal decision theorist.

Evidential decision theory comes in many flavours. On one account of evidential decision theory, it can be defined as the claim that it is not probabilities such as $p(X \mathbin{\square\!\!\rightarrow} Y)$ that should guide one's decision, but rather probabilities such as $p((X \mathbin{\square\!\!\rightarrow} Y) \mid X)$. That is, instead of asking yourself, "what is the probability that if I were to do X, then Y would be the case?", a rational decision maker should ask, "what is the probability that if I were to do X, then Y would be the case given that I do X?" To see how this seemingly unimportant difference affects the analysis, it is helpful to reconsider the psychopath case. Clearly, $p(($press button $\mathbin{\square\!\!\rightarrow}$ dead$) \mid$ press button$)$ is very high, since nearly everyone who presses the button is a psychopath. For analogous reasons, $p($press button $\mathbin{\square\!\!\rightarrow}$ live in a world without psychopaths $\mid$ press button$)$, is quite low. Hence, if these probabilities are used for calculating the expected utilities of the two alternatives, it follows that it is rational not to press the button, which tallies well with our intuitive appraisal of what is required by a rational decision maker. Furthermore, in Newcomb's problem, advocates of evidential decision theory argue that it is rational to take one box. This is because the probability $p(($take one box $\mathbin{\square\!\!\rightarrow}$ \$1M$) \mid$ take one box$)$ is high, while $p(($take two boxes $\mathbin{\square\!\!\rightarrow}$ \$1M$) \mid$ take two boxes$)$ is low.

Unfortunately, evidential decision theory is not a view without downsides. A potentially powerful objection is that it seems to require that the decision maker can somehow ascribe probabilities to his or her own choices. Furthermore, according to this objection, this is incoherent because one's own choices are not the kind of things one can reasonably ascribe probabilities to. (We will shortly explain why.) The reason why one has to ascribe probabilities to one's own choices is the simple fact that $p(X|Y) = \dfrac{p(X \wedge Y)}{p(Y)}$, as explained in Chapter 6. In the psychopath case Y is the act of pressing the button, whereas X is the fact that I live. Clearly, the former is something I trigger myself. In Newcomb's problem Y is the act of taking one box, and X the monetary outcome.

So why would it be incoherent to ascribe probabilities to one's own choices? First, if probabilities are taken to be objective it seems difficult to reconcile the idea of a decision maker making a free choice with the thought that your choices are somehow governed by probabilistic processes. How can something be the outcome of a deliberative process *and* a random event? Second, similar problems arise also if probabilities are taken to be subjective and defined in terms of preferences over bets. If the decision maker assigns subjective probabilities to his own choices, this readiness to accept bets becomes an instrument for measuring his own underlying dispositions to act. For example, if you consider $10 to be a fair price for a bet in which you win $40 if the hypothesis Y = 'I will take one box in Newcomb's problem' turns out to be true, but win nothing otherwise, your subjective probability for Y is 10/40 (given that your utility of money is linear). Several authors have pointed out that this use of bets for measuring subjective probabilities for your own acts gives rise to the following problem: (1) If the decision maker ascribes subjective probabilities to his present alternatives, then he must be prepared to take on bets on which act he will eventually choose. (2) By taking on such bets, it becomes more attractive for the decision maker to perform the act he is betting on: It becomes tempting to win the bet by simply choosing the alternative that will make him win the bet. Therefore, (3) the 'measurement instrument' (that is, the bets used for eliciting subjective probabilities) will interfere with the entity being measured, that is, the decision maker's subjective degree of belief that he will choose a certain act. The conclusion will therefore be that the decision maker's subjective probability will not reflect his true preference among the bets.

However, it is not obvious that the problem with self-predicting probabilities is a genuine threat to evidential decision theory. First, the evidential decision theorist could perhaps find a way of expressing the position that does not presuppose that the decision maker is able to assign any probabilities to his own choices. Second, it could also be argued that the difficulty with self-predicting probabilities is not as severe as it appears to be. If probabilities are taken to be subjective, the betting situation could be modified such that the agent asks a well-informed friend who knows him very well to do the betting, without telling the agent whether he will win or lose the bet if the alternative he bets on is chosen. By separating the belief-generating mechanism from the deliberative mechanism in this manner, the problem outlined above would no longer arise. Another strategy for avoiding the difficulty outlined above could be to keep the stakes of the bets to a minimum, such that they become virtually negligible in comparison to the value of the outcomes. The underlying thought is that since the measurement theoretical objection presupposes that the net gain made by choosing in accordance with one's bets is of a certain size, one could block that assumption by setting a suitable limit for the bet that is relative to each decision situation.

Box 9.1   Death in Damascus

Which decision theory is best, the causal or the evidential version? A possible, but depressing, conclusion is that there are cases in which neither the causal nor the evidential theory can help us. The following example, suggested by Gibbard and Harper, supports this conclusion.

> Consider the story of the man who met death in Damascus. Death looked surprised, but then recovered his ghastly composure and said, "I am coming for you tomorrow". The terrified man that night bought a camel and rode to Aleppo. The next day, death knocked on the door of the room where he was hiding, and said "I have come for you". "But I thought you would be looking for me in Damascus", said the man. "Not at all", said death "that is why I was surprised to see you yesterday. I knew that today I was to find you in Aleppo". Now suppose the man knows the following. Death works from an appointment book which states time and place; a person dies if and only if the book correctly states in what city he will be at the stated time. The book is made up weeks in advance on the basis of highly reliable predictions. An appointment on the next day has been inscribed for him. Suppose, on this basis, the man would take his being in Damascus the next day as strong evidence that his appointment with death is in Damascus, and would take his being in Aleppo the next day as strong evidence that his appointment is in Aleppo ... If ... he decides to go to Aleppo, he then has strong grounds for expecting that Aleppo is where death already expects him to be, and hence it is rational for him to prefer staying in Damascus. Similarly, deciding to stay in Damascus would give him strong grounds for thinking that he ought to go to Aleppo ... (Gibbard and Harper [1978] 1988: 373–4)

In this case it seems that no matter which of the two cities the man decides to go to, it would have been better for him to go to the other one, and he knows this already when he makes the decision. However, note that the man is rightfully convinced that his decision is causally independent of Death's prediction. It is not the case that Death's location is caused by the man's decision to go to one of the two cities. In a case like this, decision theorists say that there is no *stable* act available. By definition, an act is stable if and only if, given that it is performed, its unconditional expected utility is maximal. Hence, since the expected utility of performing the other alternative is always higher, no matter which alternative is chosen, there is no stable alternative available for the man in Damascus.

Death in Damascus is in many respects similar to Newcomb's problem. By definition, the decision maker's choice is causally independent of the prediction, but it nevertheless turns out that the prediction is almost always correct. When presented with Death in Damascus, decision theorists tend to react in many different ways. One option is, naturally, to argue that there is something wrong with the example. Richard Jeffrey, a well-known advocate of evidential decision theory, thinks that if presented with this kind of example, "you do well to reassess your beliefs and desires before choosing" (1983: 19). That is, Jeffrey's point is that a case like this can arise only if the decision maker's desires and beliefs about the world is somehow incorrect or pathological.

Another option is to argue that a case like Death in Damascus is in fact a genuine decision problem that could eventually materialise, although it does not follow that some alternative must therefore be rationally permissible. It could simply be the case that in a situation like this no alternative is rational. No matter what the man decides to do, he will fail to act rationally. This is because he unfortunately faces a decision problem that cannot be properly analysed by any theory of rationality.

A third option, suggested by Harper (1986), is to argue that the lack of a stable alternative is due to the fact we have not yet considered any mixed acts. As explained in Chapter 3, a mixed act is an act generated by randomising among the acts in the initial set of alternatives. For example, if the man decides to toss a coin and go to Damascus if and only if it lands, say, heads up, and we stipulate that Death can predict only *that* the man will toss a coin but not the outcome of the toss, then it no longer follows that it would have been better to choose some other alternative. Hence, we have identified a stable alternative. That said, one could of course reformulate the example by simply excluding the possibility of choosing a mixed alternative. Imagine, for example, that the man is told by Death that if he performs a mixed act Death will invariably notice that and torture him to death.

Exercises

9.1 Imagine that Newcomb's problem is reformulated such that box B_1 contains \$0 instead of \$1,000. Then this would no longer be a decision problem in which two major decision principles seem to come into conflict. Explain!

9.2 Newcomb's problem has little to do with the principle of maximising expected utility. Propose an alternative rule for decision making under risk, and show that it comes into conflict with the dominance principle in Newcomb-style cases.

9.3 (a) Suppose that $p(X \mathbin{\square\!\!\rightarrow} Y) = p(\neg X \mathbin{\square\!\!\rightarrow} Y) = 0$. What can you then conclude about Y?

(b) Suppose that $p(X \mathbin{\square\!\!\rightarrow} Y) = p(\neg X \mathbin{\square\!\!\rightarrow} Y) = 1$. What can you conclude about Y?

9.4 In the Paul-and-the-psychopath case we stipulated that the number of psychopaths in the world was known to be small. Why?

9.5 (a) Suppose that $p((X \mathbin{\square\!\!\rightarrow} Y) \mid X) = p((\neg X \mathbin{\square\!\!\rightarrow} Y) \mid \neg X) = 0$. What can you then conclude about Y?

(b) Suppose that $p((X \mathbin{\square\!\!\rightarrow} Y) \mid X) = p((\neg X \mathbin{\square\!\!\rightarrow} Y) \mid X) = 1$. What can you conclude about Y?

9.6 Death offers you a choice between \$50 and \$100. He adds that if you make a *rational* choice, he will kill you. What should you do? (You think \$100 is better than \$50, and you have no desire to die.)

Solutions

9.1 The dominance principle would no longer give the decision maker a reason to take both boxes. Hence, we would no longer face an apparent clash between the dominance principle and the principle of maximising expected utility.

9.2 Here is an example: Suppose you think improbable outcomes should be assigned less weight, proportionally speaking, than very probable outcomes. Then a similar conflict between dominance and the new rule would of course arise again.

9.3 (a) Y will not occur, no matter whether you do X or $\neg X$.

(b) Y will occur, no matter whether you do X or $\neg X$.

9.4 Otherwise the temptation to press the button would not give him new evidence for thinking that he is himself a psychopath. If nearly everyone is a psychopath, it is likely that he is one too; hence, he should not press the button in the first instance.

9.5 (a) No matter whether you were to do X, Y would not be the case no matter whether you did X or not.

(b) No matter whether you were to do X, Y would be the case no matter whether you did X or not.

9.6 If your choice is rational, Death will punish you. This indicates that you should try to do what you know is irrational. But can it really hold true that you *should* do what you know to be irrational?

10 Bayesian vs. non-Bayesian decision theory

The term *Bayesianism* appears frequently in books on decision theory. However, it is surprisingly difficult to give a precise definition of what Bayesianism is. The term has several different but interconnected meanings, and decision theorists use it in many different ways. To some extent, Bayesianism is for decision theorists (but not all academics) what democracy is for politicians: Nearly everyone agrees that it is something good, although there is little agreement on what exactly it means, and why it is good. This chapter aims at demystifying the debate over Bayesianism. Briefly put, we shall do two things. First, we shall give a rough characterisation of Bayesian decision theory. Second, we shall ask whether one can give any rational argument for or against Bayesianism.

10.1 What is Bayesianism?

There are almost as many definitions of Bayesianism as there are decision theorists. To start with, consider the following broad definition suggested by Bradley, who is himself a Bayesian.

> Bayesian decision theories are formal theories of rational agency: they aim to tell us both what the properties of a rational state of mind are … and what action it is rational for an agent to perform, given the state of mind … (Bradley 2007: 233)

According to this definition, Bayesianism has two distinct components. The first tells us what your state of mind ought to be like, whilst the second tells us how you ought to act given that state of mind. Let us call the two components the *epistemic* and the *deliberative* component, respectively. The epistemic component of Bayesianism is a claim about what rational agents ought to believe, and which combinations of beliefs and desires are rationally permissible. In essence, the theory holds that one is free to believe whatever one wishes *as long as* one's beliefs can be represented by a subjective probability function, and those beliefs are updated in accordance with Bayes' theorem (see Chapter 6). This means that Bayesian epistemology offers virtually no substantial advice on *how* one ought to go about when exploring the world. The theory merely provides a set of structural restrictions on what it is permissible to believe, and how one should be permitted to revise those beliefs in light of new information. Bayesians thus maintain that all beliefs come in degrees, and that the theory of subjective probability provides an accurate account of how degrees of belief ought to be revised. (Some Bayesians advocate objective versions of the theory, according to which probabilities are objective features of the world such as relative frequencies or physical propensities. Naturally, the mathematical framework for how probabilities ought to be updated is the same.)

Since the correct way to update one's degrees of beliefs is to apply Bayes' theorem, discussions of Bayesian epistemology are apt for formalisation. The purely mathematical part of Bayesian epistemology is impeccable. Bayes' theorem, and every principle derived from it, are mathematical results that it makes little sense to question. However, a major issue of disagreement among scholars working on Bayesian epistemology is how to determine the 'prior probabilities' needed for getting the Bayesian machinery off the ground (see Chapter 6). What is, for example, the prior probability that a newly discovered disease is contagious, i.e. what is the probability that it is contagious before we have started collecting any further information? As explained in Chapter 6, an influential strategy for tackling this problem is to prove that even people with different priors at the beginning of a series of experiments will come to agree in the end, after sufficiently many iterations.

The deliberative component of Bayesianism is supposed to tell us what action is rational for the agent to perform, given his or her present (and hopefully perfectly rational) state of mind. The principles below summarise the Bayesian account of rational action. Ramsey, Savage and many contemporary decision theories endorse all three conditions.

1. Subjective degrees of belief can be represented by a probability function defined in terms of the decision maker's preferences over uncertain prospects.

2. Degrees of desire can be represented by a utility function defined in the same way, that is, in terms of preferences over uncertain prospects.
3. Rational decision makers act *as if* they maximise subjective expected utility.

Principles 1 and 2 are discussed in detail in Chapters 5 and 7. However, principle 3 is new. Exactly what does it mean to say that rational decision makers act *as if* they maximise subjective expected utility? A brief answer is that the decision maker does *not* prefer an uncertain prospect to another *because* she judges the utilities and probabilities of the outcomes to be more favourable than those of another. Instead, the well-organised structure of the agent's preferences over uncertain prospects logically implies that the agent can be described *as if* her choices were governed by a utility function and a subjective probability function, constructed such that a prospect preferred by the agent always has a higher expected utility than a non-preferred prospect. Put in slightly different words, the probability and utility functions are established by reasoning backwards: since the agent preferred some uncertain prospects to others, and the preferences over uncertain prospects satisfy a number of structural axioms, the agent behaves as if she had acted from a subjective probability function and a utility function, both of which are jointly consistent with the principle of maximising expected utility. Thus, the axiomatic conditions on preferences employed in Bayesian theories merely restrict which *combinations* of preferences are legitimate. Blackburn has expressed this point very clearly by pointing out that Bayesian theories merely constitute a "grid for imposing interpretation: a mathematical structure, designed to render processes of deliberation mathematically tractable, whatever those processes are" (Blackburn 1998: 135).

The most fundamental Bayesian preference axiom is the ordering axiom. It holds that for any two uncertain prospects, the decision maker must be able to state a clear and unambiguous preference, and all such preferences must be asymmetric and transitive. However, the ordering axiom does not tell the agent whether she should, say, prefer red wine to white. The axiom only tells the agent that *if* red wine is preferred to white, *then* she must not also prefer white to red. This is what Blackburn means with his remark that Bayesianism can be conceived of as a grid that will "render processes of deliberation mathematically tractable, whatever those processes are". So briefly put, Bayesians argue that subjective probabilities and utilities can be established by asking agents to state preferences over uncertain prospects.

Preferences are, according to the mainstream view, revealed in choice behaviour. If the agent is offered a choice between two uncertain prospects and chooses one of them, it is reasonable to conclude that he preferred the chosen one. (Some Bayesians go as far as saying that preferences could be *identified* with choices.) The following example illustrates how information about preferences can be utilised for eliciting subjective probabilities and utilities by reasoning backwards: suppose that you wish to measure your subjective probability that your new designer watch, worth $1,000, will get stolen. If you consider $50 to be a fair price for insuring the watch – that is, if that amount is the highest price you are willing to pay for a bet in which you win $1,000 if the watch gets stolen, and nothing otherwise, then your subjective probability is *approximately* $\frac{50}{1,000} = 0.05$. Of course, in order to render this approximation precise a linear measure of utility must be established, since most agents have a decreasing marginal utility for money. However, had one known the probability of at least one event, a utility function could have been extracted from preferences over uncertain prospects in the same way as with subjective probabilities. To see this, suppose that we somehow know that a person named Baker considers the probability of 'rain in Cambridge today' to be fifty per cent. Also suppose that he prefers a gold watch to steel watch, to a plastic watch. Then, if Baker is indifferent between the prospect of getting a steel watch for certain, and the prospect of getting the gold watch if the state 'rain in Cambridge today' obtains, and the plastic watch if it does not, then Baker's utility for the three possible outcomes can be represented on a linear scale in the following way: A gold watch is worth 1, a steel watch is worth 0.5 and a plastic watch is worth 0 (because then the expected utility of the two prospects will be equal).

The watch example serves as an illustration of the fact that Bayesians can extract subjective probabilities *and* utilities from preferences over uncertain prospects, *given* that it can somehow be established that 'rain in Cambridge today' (Event 1) and 'no rain in Cambridge today' (Event 2) are considered equally likely. In order to do this, Bayesians use the following trick, first discovered by Ramsey: Suppose that Baker strictly prefers one object – say, a fancy designer watch – to another object. Then, if Baker is indifferent between the prospect in which (i) he wins the first object if Event 1 occurs and the second object if Event 2 occurs, and the prospect in which (ii) he wins the second object if Event 1 occurs and the first object if Event 2 occurs, then the two events are by definition equally probable. A simple numerical example can help clarify this somewhat technical point. Imagine that the agent considers the mutually exclusive events R and $\neg R$ to be equally probable. Then he will be indifferent between winning, say, (i) 200 units of utility if R occurs and 100 units if $\neg R$ occurs, and (ii) 100 units if R occurs and 200 units if $\neg R$ occurs. This holds true, no matter what his attitude to risk is. All that is being assumed is that the agent's preference is entirely fixed by his beliefs and desires.

10.2 Arguments for and against Bayesianism

A major reason for accepting a Bayesian approach is that it provides a *unified* answer to many important questions in decision theory, epistemology, probability theory and statistics. If you convert to Bayesianism, the answer to nearly all questions you are interested in will always be based on the same principles, viz. principles (1)–(3) as outlined above. This intellectual economy certainly contributes towards making your theoretical framework rather elegant. Furthermore, in addition to being elegant, Bayesianism also avoids a number of potentially hard metaphysical questions. There is no need to worry about what the external world really is like. All that matters is our beliefs and desires. For example, even if you think the world is deterministic and that chance does not exist, you can still be a Bayesian, since probabilities on this view ultimately refer to degrees of belief. If you wish to advocate some alternative interpretation of probability, you cannot avoid making substantial metaphysical assumptions about the nature of chance, which may prove difficult to justify.

Another important reason for seriously considering the Bayesian approach is the high degree of precision it offers. Many issues about human choice behaviour and problems in epistemology can be treated in very precise ways by applying Bayesian techniques. In principle, nearly everything can be represented by numbers, and the Bayesian can also tell you how those numbers should be updated in light of new information. That said, the critics of Bayesianism complain that the precision the numbers give us is false or unjustified. The subjective probabilities and utilities do not refer to any existing properties in the real world, because no actually existing human being is likely to satisfy all the structural principles imposed upon us by the theory. However, a possible reply to this is that Bayesianism is a theoretical ideal, i.e. something we should strive to achieve, rather than a description of actually existing people. Arguably, it might be fruitful to strive to achieve a theoretical ideal, even if we know that we will never *fully* achieve it.

That said, there is also another objection to Bayesian decision theory, which is potentially a much bigger threat. Strictly speaking, when Bayesians claim that rational decision makers behave 'as if' they act from a subjective probability function and a utility function, and maximise subjective expected utility, this is merely meant to prove that the agent's preferences over alternative acts can be *described* by a representation theorem and a corresponding uniqueness theorem, both of which have certain technical properties. (See Chapter 7.) Bayesians do not claim that the probability and utility functions constitute genuine *reasons* for choosing one alternative over another. This seems to indicate that Bayesian decision theory cannot offer decision makers any genuine action guidance. Let us take a closer look at this objection.

Briefly put, the objection holds that Bayesians 'put the cart before the horse' from the point of view of the deliberating agent. A decision maker who is able to state a complete preference ordering over uncertain prospects, as required by Bayesian theories, already knows what to do. Therefore, Bayesian decision makers do not get any new, action-guiding information from their theories. So roughly put, the argument seeks to show that even if all the axiomatic constraints on preferences proposed by Bayesians were perfectly acceptable from a normative point of view, the Bayesian representation and uniqueness theorems would provide no direct action guidance. This is because the agent who is about to choose among a large number of very complex acts must know already from the beginning which act(s) to prefer. This follows directly from the ordering axiom for uncertain prospects, which in one version or another is accepted by all Bayesians. For the deliberating agent, the output of a Bayesian decision theory is thus not a set of preferences over alternative acts – these preferences are on the contrary used as input to the theory. Instead, the output of a Bayesian decision theory is a (set of) probability and utility function(s) that can be used to describe the agent as an expected utility maximiser. This is why ideal agents can only be described *as if* they were acting from this principle.

In order to spell out the argument in more detail, it is helpful to imagine two decision makers, A and B, who have exactly the same beliefs and desires. They like and dislike the same books, the same wines, the same restaurants, etc., and they hold the same beliefs about past, present and future events. For instance, they both believe (to exactly the same degree) that it will rain tomorrow, and they dislike this equally much. However, as always in these kind of examples, there is one important difference: A is able to express his preferences over uncertain prospects in the way required by Bayesians. That is, for a very large set of risky acts {x, y, ...}, decision maker A knows for sure whether he prefers x to y, or y to x or is indifferent between them. Furthermore, his preferences conform with the rest of the Bayesian preference axioms as well. But agent B is more like the rest of us, so in most cases he does not know if he prefers x to y. However, B's inability to express preferences over alternative acts is not due to any odd structure of his beliefs and desires; rather, it is just a matter of his low capacity to process large amounts of information (so his preferences conform with the Bayesian axioms in an implicit sense). Since B's tastes and beliefs are exactly parallel to those of A, it follows that in every decision, B ought to behave as A would have behaved; that is, B can read the behaviour of A as a guidebook for himself.

Decision maker A is designed to be the kind of highly idealised rational person described by Ramsey, Savage and contemporary Bayesians. Suppose that you are A and then ask yourself what output you get from normative decision theory as it is presented by Bayesians. Do you get any advice about what acts to prefer, i.e. does this theory provide you with any action guidance? The answer is no. On the contrary, even if A was to decide among a large number of very complex acts, it is assumed in the ordering axiom that A knows already from the beginning which act(s) to prefer. So for A, the output of decision theory is not a set of preferences over risky acts – these preferences are, rather, used as input to the theory. Instead, as shown above, the output of a decision theory based on the Bayesian approach is a (set of) probability and utility function(s) that can be used to describe A as an expected utility maximiser. Again, this is why ideal agents do not prefer an act *because* its expected utility is favourable; they can only be described *as if* they were acting from this principle. So the point is that Bayesians take too much for granted. What Bayesians use as input data to their theories is exactly what a decision theorist would like to obtain as output. In that sense, theories based on the Bayesian approach 'put the cart before the horse' from the point of view of the deliberating agent.

Of course, Bayesians are not unaware of this objection. To understand how they respond, we need to bring the non-ideal decision maker B into the discussion. Recall that for B the situation is somewhat different. He does not yet know his preference between (all) acts. Now, a Bayesian theory has *strictly* speaking nothing to tell B about how to behave. However, despite this, Bayesians frequently argue that a representation theorem and its corresponding uniqueness theorem are normatively relevant for a non-ideal agent in indirect ways. Suppose, for instance, that B has access to *some* of his preferences over uncertain prospects, but not all, and also assume that he has partial information about his utility and probability functions. Then, the Bayesian representation theorems can, it is sometimes suggested, be put to work to 'fill the missing gaps' of a preference ordering, utility function and probability function, by using the initially incomplete information to reason back and forth, thereby making the preference ordering and the functions less incomplete. In this process, some preferences for risky acts might be found to be inconsistent with the initial preference ordering, and for this reason be ruled out as illegitimate.

That said, this indirect use of Bayesian decision theory seems to suffer from at least two weaknesses. First, it is perhaps too optimistic to assume that the decision maker's initial information always is sufficiently rich to allow him to fill *all* the gaps in the preference ordering. Nothing excludes that the initial preferences over uncertain prospects only allow the decision maker to derive the parts of the utility and probability function that were already known. The second weakness is that even if the initial information happens to be sufficiently rich to fill the gaps, this manoeuvre offers no theoretical *justification* for the initial preference ordering over risky acts. Why should the initial preferences be retained? In *virtue of what* is it more reasonable to prefer x to y than to prefer y to x? Bayesians have little to say about this. On their view, the choice between x and y is ultimately a matter of taste, even if x and y are very risky acts. As long as no preference violates the structural constraints stipulated by the Bayesian theory, everything is fine. But is not this view of practical rationality a bit too uninformative? Most people would surely agree that decision theorists ought to be able to say *much more* about the choice between *x* and *y*. An obvious way to address this problem would be to say that rationality requires more than just coherent preferences. The axioms are necessary but perhaps not sufficient conditions for rationality. A rational decision maker must somehow acknowledge probabilities and utilities measured but not defined by the the Bayesian theory. However, an extended theory that meets this requirement would no longer be purely Bayesian.

10.3 Non-Bayesian approaches

Non-Bayesian decision theories can be divided into two subclasses, viz. *externalist* and *internalist* theories, respectively. Externalists argue that an act is rational not merely by virtue of what the decision maker believes and desires. This set of theories thus rejects the so-called Humean belief–desire account of practical rationality, according to which acts can be interpreted and rationalised by identifying the beliefs and desires that prompted the decision maker to perform the act in question. From an externalist perspective, the belief–desire model is too narrow. Rationality is also (at least partly) constituted by facts about the external world. For example, theories advocating objective concepts of probability, such as Laplace, Keynes and Popper, are all non-Bayesian in an externalist sense. According to their view, the kind of probability that is relevant when making decisions is not the decision maker's subjective degree of belief. Instead, the concept of probability should rather be interpreted along one of the objective lines discussed in Chapter 7. Whether some form of externalism about practical rationality should be accepted is ultimately a matter of what we take a normative reason to be. Is a normative reason constituted solely by beliefs and desires, or is some 'external' component required? It is beyond the scope of this book to assess the debate on this apparently difficult question.

So what about the other, purely internalist alternative to Bayesianism? Briefly put, internalists agree with Bayesians that the decision maker's beliefs and desires are all that matters when adjudicating whether an act is rational or irrational. However, unlike Bayesians, the non-Bayesian internalist denies that rational decision makers merely act 'as if' they were acting from the principle of maximising expected utility. The non-Bayesian internalist claims that rational decision makers choose an act over another *because* its subjective expected utility is optimal. One's beliefs and desires with respect to the risky acts *x* and *y* thus constitute reasons for preferring one act to another. That is, preferences over risky acts are not just a tool that is used for *measuring* degrees of belief and desire. Instead, one's beliefs and desires constitute reasons for preferring one risky act over another.

Is non-Bayesian internalism a tenable alternative to Bayesianism? In recent years, decision theorists have paid surprisingly little attention to this approach. However, during the golden era of axiomatic decision theory, in the 1950s and 1960s, a number of decision theorists proposed views that are best conceived of as genuine alternatives to the Bayesian way of aggregating beliefs and desires. For example, in the discussion on subjective probability in Chapter 7, we briefly mentioned DeGroot's axiomatisation of subjective probability. Recall that DeGroot's theory begins with the assumption that agents can make *qualitative* comparisons of whether one event is more likely to occur than another. DeGroot thereafter showed that if the decision maker's qualitative judgements are sufficiently fine-grained and satisfy a number of structural conditions, then there exists a function p that assigns real numbers between 0 and 1 to all events, such that one event is judged to be more likely than another if and only if it is assigned a higher number, and this function p satisfies the axioms of the probability calculus. Thus, on DeGroot's view, subjective probabilities are obtained by fine-tuning qualitative data, thereby making them quantitative. The probabilistic information is present from the beginning, but after putting qualitative information to work the theory becomes quantitative. Since subjective probabilities are not obtained by asking the decision maker to state a set of preferences over risky acts, this theory is not a Bayesian theory in the sense stipulated above.

Naturally, decision theorists have also proposed non-Bayesian accounts of utility. Luce's probabilistic theory of utility is perhaps the most interesting account, which is still widely discussed in the literature. As explained in Chapter 5, Luce managed to show that we can ascribe a utility function to a decision maker given that his choice behaviour is probabilistic, and conforms to the choice axiom. Even though this account of utility clearly invokes a number of rather strong philosophical assumptions, it clearly counts as a non-Bayesian account of utility: the utility function is not derived from a set of preferences over risky alternatives. Hence, it seems that a utility function derived in this way could be put to work for explaining *why* one risky alternative ought to be preferred to another, given that subjective probability is interpreted along the lines proposed by DeGroot.

Of course, it might be thought that the difference between the various accounts of practical rationality considered here are rather minuscule and not really worth that much attention. However, even if the discussion may appear a bit theoretical (which it, of course, is) the controversy over Bayesianism is likely to have important practical implications. For example, a forty-year-old woman seeking advice about whether to, say, divorce her husband, is likely to get very different answers from the Bayesian and his critics. The Bayesian will advise the woman to first figure out what her preferences are over a very large set of risky acts, including the one she is thinking about performing, and then just make sure that all preferences are consistent with certain structural requirements. Then, as long as none of the structural requirements is violated, the woman is free to do whatever she likes, no matter what her beliefs and desires actually are. The non-Bayesian externalist would advise her to consider the objective probability that she will be happy if she divorces her husband, whereas the non-Bayesian internalist will advise the woman to first assign numerical utilities and probabilities to her desires and beliefs, and then aggregate them into a decision by applying the principle of maximising expected utility.

Exercises

10.1 Explain, by using your own words, the term 'Bayesianism'.

10.2 Your subjective degree of belief in B is 0.1 and your degree of belief in A given B is 0.4. You then come to believe that the unconditional probability of A is 0.90. To what degree should you believe B given A? (That is, what is $p(B|A)$?)

10.3 Pete is a Bayesian. He tells you that he prefers x to z, and the lottery ypw to xpw. He thereafter asks you to advise him whether he should prefer y to z, or z to y, or be indifferent. What do you tell him?

10.4 Suppose that Pete in Exercise 10.3 prefers z to y. His preferences are then incoherent. Can you advise him which of his preferences he ought to revise?

10.5 In the account of Bayesianism given here, we identified an epistemic and a deliberative component, respectively. But is there really any fundamental difference between choosing what to believe and choosing what to do – perhaps choosing a belief is a way of performing an act? If so, do you think it would be possible to reduce Bayesianism to a single, unified theory about rational deliberation?

Solutions

10.1 See the discussion in the first pages of the chapter.

10.2 We know from Chapter 6 that $p(B|A) = \dfrac{p(B) \cdot p(A|B)}{p(A)}$. Hence, $p(B|A) =$ 0.044.

10.3 He must not prefer z to y, because then his preferences would be incoherent.

10.4 No, he is free to revise any one of his preferences (thereby making them coherent). It would be incorrect to think that there is some specific preference that he must revise.

10.5 I leave it to the reader to make up her mind about this.

11 Game theory I: Basic concepts and zero-sum games

Game theory studies decisions in which the outcome depends partly on what other people do, and in which this is known to be the case by each decision maker. Chess is a paradigmatic example. Before I make a move, I always carefully consider what my opponent's best response will be, and if the opponent can respond by doing something that will force a checkmate, she can be fairly certain that I will do my best to avoid that move. Both I and my opponent know all this, and this assumption of *common knowledge of rationality* (CKR) determines which move I will eventually choose, as well as how my opponent will respond. Thus, I do not consider the move to be made by my opponent to be a state of nature that occurs with a fixed probability independently of what I do. On the contrary, the move I make effectively decides my opponent's next move.

Chess is, however, not the best game to study for newcomers to game theory. This is because it is such a complex game with many possible moves. Like other parlour games, such as bridge, monopoly and poker, chess is also of limited practical significance. In this chapter we shall focus on other games, which are easier to analyse but nevertheless of significant practical importance. Consider, for example, two hypothetical supermarket chains, Row and Col. Both have to decide whether to set prices high and thereby try to make a good profit from every item sold, or go for low prices and make their profits from selling much larger quantities. Naturally, each company's profit depends on whether *the other company* decides to set its prices high or low.

If both companies sell their goods at high prices, they will both make a healthy profit of $100,000. However, if one company goes for low prices and the other for high prices, the company retailing for high prices will sell just enough to cover their expenses and thus make no profit at all ($0), whereas the other company will sell much larger quantities and make $120,000.

Table 11.1

		Col	
		High prices	Low prices
Row	High prices	$100,000, $100,000	$0, $120,000
	Low prices	$120,000, $0	$10,000, $10,000

Furthermore, if both companies set their prices low each of them will sell equally much but make a profit of only $10,000. Given that both companies wish to maximise their profit, what would you advise them to do? Consider the game matrix in Table 11.1.

Player Row is essentially choosing which row to play, whereas Col is deciding which column to play. The row and column chosen by them together uniquely determine the outcome for both players. In each box describing the outcomes, the first entry refers to the profit made by Row and the second entry to that made by Col.

In order to advise Row what to do, suppose that we somehow knew that Col was going to keep its prices high. Then it would clearly follow that Row should go for low prices, since a profit of $120,000 is better than one of $100,000. Now consider what happens if we instead suppose that Col is going to set its prices low. In that case, Row is effectively making a choice between either making $0 (if it goes for high prices), or making $10,000 (if it goes for low prices). Of course, in that case it would also be better to go for low prices. Hence, we can conclude that it is rational for Row to go for low prices *no matter what Col decides to do*.

Now look at the game from Col's perspective. If Row chooses high prices, it would be better for Col to choose low prices, because making a profit of $120,000 is better than making 'only' $100,000. Furthermore, if Row were to decide on setting its prices low, then it would also be better for Col to set its prices low, since $10,000 is more than $0. Hence, we can conclude that it is rational for Col to also set its prices low *no matter what Row decides to do*.

Since it is rational for *both* players to set their prices low, regardless of what the other player decides to do, it follows that each company will end up making a profit of only $10,000. This is a somewhat surprising conclusion. By looking at the matrix above, we can clearly see that there is another outcome in which *both* companies would make much more money: if *both* companies were to decide on setting their prices high, each company would make a healthy profit of $100,000. However, if both companies act rationally, then they cannot reach this optimal outcome, even though it would be better for both of them. This is indeed a surprising conclusion, which is also of significant practical importance. For instance, this example explains why companies often try to agree in secret to set prices high, although government authorities and consumer organisations do their best to prevent such deals: if big retailers are allowed to communicate about price levels, and can somehow form binding agreements about price levels, then the retailers will make a big profit at the consumers' expense. In effect, many countries have implemented legal rules prohibiting cartels.

11.1 The prisoner's dilemma

The supermarket-game is an example of a much larger class of games, known as *prisoner's dilemmas*. Before we explain more in detail how games can be classified as belonging to different classes, we shall take a closer look at this important example.

Imagine that the police have caught two drug dealers, Row and Col. They are now sitting in two separate cells in the police station. Naturally, they cannot communicate with each other. The prosecutor tells the prisoners that they have one hour to decide whether to confess their crimes or deny the charges. The prosecutor, who took a course in game theory at university, also informs the prisoners that there is not enough evidence for convicting them for all their offences unless at least one of them decides to confess. More precisely put, the legal situation is as follows: If both prisoners confess, they will get ten years each. However, if one confesses and the other does not, then the prisoner who confesses will be rewarded and get away with just one year in prison, whereas the other one will get twenty years. Furthermore, if both prisoners deny the charges they will nevertheless be sentenced to two years for a series of well-documented traffic offences. The matrix in Table 11.2 summarises the strategies and outcomes of the game.

To analyse this game, we first look at the matrix from Row's point of view. Row can choose between confessing or denying the charges. Suppose that Row knew that Col was going to confess. Then, Row would essentially be making a choice between ten or twenty years in prison. Thus,

Table 11.2

		Col	
		Confess	*Do not*
Row	*Confess*	−10, −10	−1, −20
	Do not	−20, −1	−2, −2

in that case it would clearly be better to confess and get ten years. Furthermore, if Row knew that Col was going to deny the charges, Row would in effect be making a decision between getting either one or two years in prison. In that case it would, of course, be better to confess and get one year instead of two. Hence, Row should confess *no matter what Col decides to do*.

Now look at the matrix from Col's point of view. There are two slightly different ways in which Col can reason, both of which are valid and lead to the same conclusion. First, Col could assume that Row will either confess or not. Then, since the outcome for Col will be better no matter what Row decides to do – since ten years is better than twenty, and one is better than two – it follows that Col should confess no matter what Row decides to do. Hence, both players will confess and end up with ten years in prison each. This holds true, despite the fact that both prisoners would get just two years had they decided not to confess.

The second way in which Col could reason is the following. Instead of considering his best response to all possible strategies made by Row, Col could simply start from the assumption that Row is rational (common knowledge of rationality, CKR). As we exemplified above, if Row is rational he will confess. Hence, if Col knows that Row is rational he *also* knows that Row will confess. Hence, there is no need to consider the possibility that Row may not confess. Moreover, since Col's best response to Row's strategy is to confess, we can conclude that both players will confess. Interestingly enough, the fact that both players are rational force them into an outcome that is *worse* for both of them.

The prisoner's dilemma shows that what is optimal for each individual need *not* coincide with what is optimal for the group. Individual rationality sometimes comes into conflict with group rationality. In prisoner's dilemmas, rationality forces both players individually to choose an outcome that is worse, for both of them, than the outcome they would have agreed on had they acted as a group seeking to minimise the total number of years in prison. In this case, both of them get ten years instead of two.

It could be objected that a prisoner who is about to confess must consider the risk of being punished by the criminal friends of the betrayed prisoner. However, note that this objection is not compatible with the assumptions we have made. As the example is set up, the number of years in prison is all that matters for the prisoners. If we take other possible consequences into account, we are in effect playing a different game, which is less interesting from a theoretical point of view. It is not the story about the prisoners that has made the prisoner's dilemma famous but rather the robust mathematical structure of the game.

So how robust is the prisoner's dilemma? Is there any way in which rational prisoners could avoid confessing? To start with, suppose that you doubt that the other prisoner is fully rational. Perhaps he will deny the charges, even though you *know* that doing so would be foolish. Does this change the situation? No. Because it is better for Row to confess, regardless of what he thinks Col will do. Hence, as long as Row is rational, he will confess. Furthermore, if Col is rational but feels uncertain about whether Row is rational, it is nevertheless rational for Col to confess. Hence, both players should confess no matter what they believe about their opponent's capability to choose rationally. Hence, in this particular game (unlike in some other games to be discussed later on), we can figure out what each rational player will do without making any strong assumptions about the opponent's degree of rationality.

Let us now consider what happens if we assume the prisoners are somehow able to communicate with each other and thus coordinate their strategies. Suppose, for instance, that the prosecutor allows the prisoners to meet in a cell for an hour. The outcome of this meeting will, of course, be that Col and Row promise each other to deny the charges, since this will render them just two years in prison. The problem is, however, that both players are rational. Therefore, when the meeting is over and Row is asked to make his choice he will reason as follows:

> My opponent Col has promised me that he will deny the charges. I am a rational person, but not a moral one, so I care only about the outcome for myself. Hence, the fact that I have promised him to deny the charges is irrelevant. If I confess and he does not, I will get just one year, which is better than two. Furthermore, it is also better to confess if Col cheats on

me and confesses. Therefore, it is better for me to confess no matter what Col eventually decides to do.

Since the game is symmetric, Col will of course reason exactly like Row and confess. Hence, our initial assumption that the players were not able to communicate about their plans was in fact superfluous. Rational prisoners will confess even if they get the opportunity to coordinate their strategies. This holds true no matter whether they trust that the other player will respect their agreement.

The prisoner's dilemma is much more than just an odd and somewhat artificial example. Prisoner's dilemmas occur everywhere in society. The supermarket-game serves as a helpful illustration. For another example, imagine a number of chemical industries located next to a beautiful river. Each firm has to decide whether to limit its emissions of pollutants into the river, and thereby make a slightly lower profit, or ignore all environmental concerns and maximise its profit. Furthermore, since the local environmental authority has too little funding, each firm knows that it is virtually risk-free to pollute. If a firm emits all its pollutants into the river, it is very unlikely that it will be punished for it. Furthermore, whether the river stays clean or not does not depend on what any single firm decides to do. The river will get dirty if and only if the large majority of firms decide to pollute. Now, the decision to be taken by each firm can be illustrated in Table 11.3, in which only the outcomes for Row are indicated.

As can be clearly seen in this matrix, it is better for Row to pollute no matter what it thinks other firms will do. A large profit and a dirty river is better than a small profit and a dirty river; and a large profit and a clean river is better than a small profit and a clean river. Furthermore, this holds true for all firms. Hence, it is rational for all firms to pollute, and to some degree this explains why so many rivers are so dirty, especially in countries with weak environmental legislation.

Note that the conclusion that it is better to pollute holds true even if all firms agree that it would be better, from each firm's point of view, to make a

Table 11.3

	Other firms pollute	They do not
My firm pollutes	Large profit and dirty river	Large profit and clean river
It does not	Small profit and dirty river	Small profit and clean river

Table 11.4

	Others use cars	They use public transport
I take my car	Comfort and traffic jam	Comfort and no traffic jam
I travel by public transportation	Less comfort and traffic jam	Less comfort and no traffic jam

slightly smaller profit and keep the river clean, thereby enabling the families of the board members to use it for recreation. Rationality prevents the firms from not polluting, even though it would be better for everyone.

Here is yet another example of a prisoner's dilemma. The citizens of a large city have to choose between getting to work by either taking their own car or by using public transportation. Taking the car is more comfortable than travelling by public transportation; however, if everyone decides to take the car there will be a huge traffic jam. (See Table 11.4.)

Look at Table 11.4 from Row's point of view. As you can see, it is better for Row to take the car no matter what she thinks the other players will do, because more comfort is better than less and whether there will be a traffic jam depends on what others do. However, this holds true for everyone; the identity of Row is irrelevant. Hence, it is rational for everyone to take their own car, rather than travelling by public transportation. The outcome of this is likely to be well known to most readers of this book: Everyone drives their own car, and this creates huge traffic jams in most cities around the world. To travel by public transportation is irrational, given that you care about your own comfort (and all other differences are negligible). So if you spent a lot of time in traffic jams when commuting to work by car this morning, it might be good to know that you are at least not doing anything irrational. But it is perhaps less comforting to know that there is an alternative outcome that would be better for everyone, namely to accept the slightly lower comfort on the train or bus and avoid all traffic jams. But as long as people are rational, we will never reach that outcome.

More generally speaking, prisoner's dilemmas arise whenever a game is symmetrical, in the sense that everyone is essentially facing the same strategies and outcomes, and the ranking of the outcomes is as in Table 11.5. Note that the numbers only refer to some ordinal ranking of outcomes. All that matters is that 4 is better than 3 and 2 is better than 1. In many games, it is taken for granted that utility is measured on an interval

Table 11.5

	They do X	They do not
I do X	2, 2	4, 1
I do not	1, 4	3, 3

scale, but in the prisoner's dilemma this rather strong assumption is superfluous. Since doing X – which may be any action – dominates not doing X, Row will do X. However, this holds true also for Col since the game is symmetrical, so everyone will end up doing X, and thereby reach the third best outcome. As we have seen from the examples above, it would not be possible to reach the very best outcome, 4, since that outcome can only be reached if some other player gets the worst outcome, 1. However, it would actually be possible for everyone to reach his or her second-best outcome – given that the players were irrational!

The prisoner's dilemma was first formulated in 1950 by Merill Flood and Melvin Dresher while working at the RAND corporation in Santa Monica, and formalised by Albert Tucker, a mathematician working at Princeton. In the 1970s, philosopher David Lewis pointed out that there is an interesting connection between the prisoner's dilemma and Newcomb's problem. To see this, imagine that you are one of the prisoners sitting in a cell thinking about whether to confess or not. Since you have good reason to believe that the other prisoner is also rational, and since the game is symmetric, it is probable that the other player will reason like you. Hence, it is probable that the other prisoner will confess if and only if you confess; and it is also probable that the other prisoner will deny the charges if and only if you do so.

However, while sitting alone in your cell you cannot be *entirely* sure that the other prisoner will reach the same conclusion as you. Let us say that your subjective probability is 0.95 that both of you will choose the same strategy. This means that you are now facing a decision problem under risk, which can be represented as in Table 11.6.

As in Newcomb's problem, the probability of the two states depends on what you do. If you confess it is very likely that the other player will also confess, but it you do not the other player is much more unlikely to do so. Suppose that the number of years in prison correspond exactly to your cardinal disutility of spending time in prison. Then, your expected

Table 11.6

	He confesses	He does not
You confess	−10, probability 0.95	−1, probability 0.05
You do not	−20, probability 0.05	−2, probability 0.95

utility of confessing is −9.55, whereas that of not doing so is only −2.9. Hence, the principle of maximising expected utility recommends you not to confess. Now apply the dominance principle to the matrix above. Clearly, −10 is better than −20 and −1 is better than −2, so the dominance principle recommends you to confess. As in the Newcomb problem, we have a clash between two of the most fundamental principles of rationality, the principle of maximising expected utility and the dominance principle.

Arguably, this example shows that the doors between decision theory and game theory are by no means watertight. Game theory can be conceived of as a special branch of decision theory, in which the probability of the states (that is, the course of action taken by other players) depends on what you decide to do. Of course, this dependence need not be causal. To emphasise this close relation between traditional decision theory and game theory also makes a lot of sense from a historical perspective. As you may recall from the brief summary of the history of decision theory in Chapter 1, von Neumann and Morgenstern developed their axiomatisation of the expected utility principle as an aid for studying game theory!

11.2 A taxonomy of games

The prisoner's dilemma is a *non-cooperative, simultaneous-strategy, symmetric, nonzero-sum* and *finite* game. In this section we shall explain all the italicised concepts, and a few more. To be exact, we shall explain the following distinctions:

- Zero-sum versus nonzero-sum games
- Non-cooperative versus cooperative games
- Simultaneous-move versus sequential-move games
- Games with perfect information versus games with imperfect information

- Non-symmetric versus symmetric games
- Two-person versus n-person games
- Non-iterated versus iterated games

In a *zero-sum* game you win exactly as much as your opponent(s) lose. Typical examples include casino games, and parlour games like chess. The total amount of money or units of utility is fixed no matter what happens. The prisoner's dilemma is, however, not a zero-sum game, simply because this criterion is not satisfied.

A *non-cooperative* game is not defined in terms of whether the players do actually cooperate or not. The official definition of a non-cooperative game is a game in which the players are not able to form binding agreements, because in some non-cooperative games players do in fact cooperate. However, they do so because it is in their own interest to cooperate. Consider Table 11.7, in which we stipulate that the players cannot cooperate in the sense that they first communicate about their strategies and then enforce a binding contract forcing them to adhere to the agreed strategies.

In this non-cooperative game, rational players will cooperate. This is because both players know that they will be strictly better off if they cooperate, irrespective of what the other player decides to do. However, the game is nevertheless a non-cooperative game, since there are no legal rules or other mechanisms that force players to keep their agreements. Games that are not non-cooperative are called *cooperative* games. By definition, players playing a cooperative game can agree on binding contracts that force them to respect whatever they have agreed on. In most Western societies, private companies can sometimes play cooperative games with other companies, in which the legal system guarantees that each player respects the agreements they have reached with other players. Such legal rules usually benefit all players.

Table 11.7

	Cooperate	Do not
Cooperate	5, 5	2, 1
Do not	1, 2	1, 1

A *simultaneous-move* game is a game in which the players decide on their strategies without knowing what the other player(s) will do. Hence, it is not necessary that the players choose their strategies at exactly the same point in time. All that matters is that no player is aware of what the other player(s) has decided on before making her move. Scissors, paper and stone, a game played by children in the US and Europe, is an example of a simultaneous-move game. Both players announce one of three alternative strategies, called *paper*, *stone* or *scissors*, without knowing what the other player will do. According to the rules of this game, *paper* beats *stone*, and *stone* beats *scissors*, and *scissors* beats *paper*. So if you play *stone* and I play *scissors*, you win. Of course, it would be no fun to play this game had it not been a simultaneous-move game. If you knew what I was going to play as you decided on your strategy, you would quickly get bored.

In *sequential* games the players have some (or full) information about the strategies played by the other players in earlier rounds. Chess is a paradigmatic example. For another example, imagine two car manufacturers who wish to maximise their profit by offering various discounts to their potential customers. Both companies make new decisions about their discount policies as soon as they learn about the other company's latest offers; for simplicity, we assume that the first company decides on its discounts on Mondays, Wednesdays and Fridays, whereas the other company makes its decisions on Tuesdays, Thursdays and Saturdays. For this to be a sequential game, it is not essential that each company remember all previous strategies played by the other company. It is sufficient that each player has *some* information about what the other company has done in the past.

However, the most interesting subset of the set of sequential games is those in which the players have *full* information about the strategies played by the other player(s). Games that fulfil this criterion are called games with *perfect information*. Hence, all games with perfect information are sequential games, but not all sequential games are games with perfect information. Chess is a paradigmatic example of a game with perfect information, as is the centipede game discussed in Section 11.3.

Many games studied by game theorists are *symmetric*, including most versions of the prisoner's dilemma. This is because such games are often

Table 11.8

	Strat 1	Strat 2		Strat 1	Strat 2
Strat 1	0, 1	0, 0	Strat 1	0, 1	0, 0
Strat 2	0, 0	1, 0	Strat 2	0, 0	0, 1

easier to study from a mathematical point of view. As briefly explained in the previous section, a symmetric game is a game in which all players face the same strategies and outcomes. This means that the identity of the players do not matter. In Table 11.8, the leftmost matrix illustrates a symmetric game, whereas the rightmost one illustrates an asymmetric game. Note that in the latter game, Col is in a better position than Row.

We now come to the distinction between *two-person* and *n-person* games. The games illustrated in the matrices in Table 11.8 are two-person games, simply because they are played by exactly two players. However, the players may be single individuals or large organisations such as multi-national companies. It is not the total number of people involved that counts, what matters is the number of players. An *n*-person game is a game played by an arbitrary number of players. Many games are relatively easy to analyse as long as only two players play them, but they may become significantly more difficult to analyse if played by *n* players. For instance, if you play Monopoly against five people instead of just one you have to think a bit harder before you decide which strategy to choose. Of course, since *n*-person games do not fit in traditional two-dimensional matrices, such games are also more difficult to illustrate graphically.

The next distinction is between *mixed* and *pure* strategies. Briefly put, to play a mixed strategy means to play a pure strategy with some probability between zero and one. For instance, to play a mixed strategy in the prisoner's dilemma means to confess with probability p and to deny the charges with probability $1 - p$, where $p > 0$. That is, the player chooses which pure strategy to play by using some random device. Mixed strategies are important from a theoretical point of view, because sometimes it is better to play a mixed strategy rather than a pure one. Imagine a general who is about to attack his enemy in one of two alternative places. If he believes that there is a spy in his headquarters, telling the enemy about his plans, the best thing to do may be to choose where to attack by tossing a fair coin. This guarantees that the general's decision will be a surprise for the enemy.

It is also important to note that some games can go on forever without coming to an end, whereas other games always end after a finite series of moves. The former kind of game is called an infinitely repeated game. For instance, commuters choosing between driving their own car to work or taking the bus play an infinitely repeated game. (The fact that humans live only for a finite number of days is of little interest from a purely theoretical point of view.) Other games are certain to come to an end at some point. A game of chess cannot go on forever, because the rules stipulate that if one and the same position arises three times the game is a draw, and there is only a finite number of possible positions.

The distinction between *iterated* and *non-iterated* games is also straightforward and easy to grasp. A *non-iterated* game is played only once, no matter how many strategies it comprises. An *iterated* game is played several times. However, the importance of this distinction is easy to overlook. Sometimes it makes a huge difference whether a game is played just once. For example, if the prisoner's dilemma is presented as a non-iterated game rational players cannot avoid ending up with the suboptimal outcomes described above. However, if the game is iterated many times (perhaps infinitely many times), there is actually another strategy that is better: to play *tit for tat*. To play tit for tat means to respond in each round with the same type of strategy the opponent played in the previous round. Hence, each player will cooperate if the opponent played a cooperative strategy in the previous round, but play a retaliative strategy if the opponent did not cooperate in the previous round.

11.3 Common knowledge and dominance reasoning

All game theorists agree that it would be great to have a single, unified theory that could be applied for solving all different kinds of games. Unfortunately, there is no such unified theory. So far, game theorists have only formulated theories that correctly solve specific subclasses of games such as, for example, two-person zero-sum games. However, even if there is no general solution that can be applied to all games, it is worth pointing out that many games can in fact be solved by just applying the dominance principle in a clever way.

The technique to be explained here is based on a number of technical assumptions, which are commonly accepted by most game theorists. The first assumption is that *all players are rational*, and therefore try to play strategies that best promote the objective they consider to be important. However, this does not mean that all players must be selfish. Some players may have a strong desire for performing what appears to be altruistic actions, so if a rational player pays attention to the wellbeing of others, he does so just because he believes that doing so will somehow promote his own wellbeing. The second assumption is that *all players know that the other players are rational*. Hence, it is not just merely the case that all players are rational; they also know that everybody else is rational. In fact, for theoretical purposes it is often fruitful to strengthen this assumption even further: Each player *knows that* the other player *knows that* the first player *knows that* the other player *knows that* ... the player in question is rational. Formulated in this way, this assumption is sometimes referred to as the assumption of nth-order common knowledge of rationality (nth order CKR). The third assumption is simply that the *dominance principle is a valid principle of rationality*. This assumption is so common that it is sometimes not even listed as an assumption. However, as we have seen in previous sections, it makes sense to accept this principle only in case one thinks the players' strategies are causally independent of each other, and if the worries arising in Newcomb-style problems can be taken care of in one way or another.

To *solve* a game is to figure out which strategies rational players would choose if confronted with the game in question. The chosen strategies determine the outcome of the game. An example is shown in Table 11.9. Two players have three strategies each to choose from; as usual, the first number refers to the outcome for Row, and the second number to the outcome for Col.

In this game, Row will certainly not play strategy R1, since that strategy is dominated by strategy R2. No matter which strategy Col decides on, Row will be better off if she plays R2 rather than R1. Therefore, both players

Table 11.9

	C1	C2	C3
R1	1, 3	2, 2	1, 0
R2	3, 2	3, 3	2, 7*
R3	1, 1	8, 2	1, 4

know for sure that Row will play either R2 or R3. This means that strategy C3 is an attractive strategy for Col, since she knows that she will not end up with 0. More precisely put, given that Row will not play R1, strategy C3 dominates C2 and C1. Hence, both players can conclude that Col will play strategy C3. Furthermore, since Col will play C3, Row will be better off playing strategy R2 rather than R3. Hence, if both players are rational and the assumption of CKR holds true, it follows that Col will play strategy C3 whereas Row will play strategy R2. This means that this pair of strategies, marked by an asterisk in the matrix, is a solution to the game. Hence, those strategies are what we should expect the players to choose, given our assumptions about their rationality and knowledge.

Box 11.1   Dominance reasoning with more than two players

The dominance reasoning illustrated above is applicable also to games with more than two players. Let us consider a game with *three* players, who have only two strategies to choose between: Player 1 has to choose between A or B; player 2 between F or G; and player 3 between X or Y, respectively. Since it is difficult to represent three-dimensional matrices on two-dimensional surfaces, we use a slightly different representation than before. Below, each of the $2 \times 2 \times 2$ possible outcomes are represented by a triple such that the first element represents the outcome for the first player, and so on. Suppose the outcomes are as follows:

$(A, F, X) = (1, 6, 9)$
$(A, F, Y) = (2, 5, 8)$
$(A, G, X) = (3, 9, 9)$
$(A, G, Y) = (4, 8, 7)$
$(B, F, X) = (2, 1, 3)^*$
$(B, F, Y) = (3, 2, 2)$
$(B, G, X) = (4, 0, 7)$
$(B, G, Y) = (5, 1, 2)$

In this example, the triples $(A, F, X) = (1, 6, 9)$ mean that player 1 chooses A rather than B and gets an outcome worth 1; and player 2 chooses F rather than G and gets an outcome worth 6; and player 3 chooses X rather than Y and gets an outcome worth 9. It takes little reflection to see that player 1 will choose B rather than A, no matter what the other player decides to do. For instance, if players 2 and 3 play F and X, then strategy B is better than A; and if the others play F and Y, then B is also better than A, and so on. This means that none of the sets of strategies listed on the first four lines will be chosen by the players. Now look at the remaining four lines. If player 2 knows that player 1 will do B, then player 2 can conclude that it will certainly be better to choose F rather than G, since 1 is better than 0 (which one gets if player 3 chooses X) and 2 is better than 1 (which one gets if player 3 chooses Y). Now look at the game from the perspective of player 3. Given that B and F will be chosen by players 1 and 2, respectively, it is clearly better to choose X rather than Y, since 3 is better than 2. Hence, the solution to this game is $(B, F, X) = (2, 1, 3)$, i.e. the fifth line, which has been marked by an asterisk.

It is worth noticing that the total sum of utility of (B, F, X) is just 6 units, which is actually *less* than the sum of utility realised by all other sets of strategies. This is yet another illustration of the amazing fact that collective rationality may conflict with individual rationality. What is best from a group perspective is not certain to be good for all individuals of the group. (Note that this point requires that utility can be somehow interpersonally aggregated, which is a controversial assumption rejected by many decision theorists.)

So far the dominance principle has directed us towards clear, unambiguous and uncontroversial conclusions. Unfortunately, this is not always the case. There are many games in which no strategies are dominated by the others. Consider the example in Table 11.10.

First look at Row's alternatives. Clearly, none of the three alternatives dominates any of the others. Then look at Col's alternatives. Again, none dominates the others. Hence, the dominance reasoning applied above cannot get started. This means that if we wish to solve this game we have to find some other principle or principles telling us what rational players should do.

Before we discuss alternative ways of solving games, it is worth pointing out that there are also cases in which dominance reasoning leads to unacceptable conclusions. The centipede game provides a classic illustration of this. This game is a finite and sequential two-person game, in which the dominance principle is applied for reasoning 'backwards'. Imagine that Sophie, the mother of Anne and Ben, tells her children that they will have the

Table 11.10

	C1	C2	C3
R1	1, 3	2, 4	1, 0
R2	3, 3	3, 3	0, 1
R3	4, 2	2, 0	1, 3

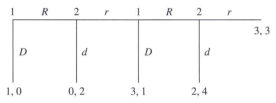

Figure 11.1

opportunity to make a series of alternating moves. Sophie then puts one
dollar in a pot on the table. In her first move, Anne must either let Sophie
add a dollar to the pot or take the dollar and quit the game. If Anne decides
to let Sophie put in a dollar, Ben must either let Sophie add one more dollar
to the pot, or stop the game by taking the two dollars on the table. If none of
them stops the game in the first round, more money will be added to the
pot; see Figure 11.1. The numbers at the bottom refer to the possible
payouts to Anne and Ben. *D* means that Anne plays down (i.e. quits the
game by grabbing the pot), whereas *d* means that Ben plays down. *R* means
that Anne plays row (i.e. adds a dollar to the pot), and *r* means that Ben plays
row. Sophie tells her children that the game will be stopped after exactly
four rounds (or a hundred; it does not matter).

Let us see what happens if the dominance principle is applied to this
game. At the last choice node, Ben has to choose between receiving either
$3 (which is the amount he gets if he plays *r*) and $4 (which he gets if
he plays d). Of course, $4 is better than $3. However, Anne knows that Ben
will reason like this already when Anne is about to make her choice at the
preceding choice node. Therefore, Anne realises that she is in fact making a
decision between receiving $3 by quitting the game, or adding an extra
dollar and getting $2. She will of course choose the first option. However,
Ben knows all this already as he makes his first move, and $2 is better than
$1, so at that point he would quit the game and take $2. Knowing all this,
Anne will quit the game already at the first choice node and take the one

dollar put on the table by her mother Sophie. Hence, Anne will get $1 whereas Ben will get nothing.

This analysis seems to be impeccable from a technical point of view, but the conclusion is nevertheless a very odd one, to say the least. Had Anne and Ben not been thinking so much about the other player's behaviour they would have ended up with $3 each. How can it be rational to get $1 or nothing, when one could have obtained $3 or more? Since decision theory is a theory of how to make rational decisions, one may suspect that there is something wrong with a theory that implies that one will *always* get $1 instead of $3 or more.

However, in response to this argument, it should be pointed out that according to the rules of the game *only irrational people* could ever get more than $1. Here is an analogy. Imagine that I promise to give you a million dollars if you decide to choose irrationally, but nothing otherwise. You want the money, so it would thus be 'rational to choose irrationally'. However, *no* theory of rationality could ever avoid this type of potential paradox! The derivation of the 'paradoxical' conclusion does not depend on any substantial principle of rationality. No matter what principle one uses for evaluating strategies, it would be 'rational to choose irrationally'. This shows at least two things. First, one may doubt whether this is a genuine problem. If so, exactly what is the problem? Second, the analogy with the million-dollar promise indicates that the alleged paradox is not directly related to the centipede game. As noted above, the problem arises because of the rules of the game, not because of the principles used for evaluating alternative strategies. In fact, the dominance principle is not essential to the centipede game. Analogous conclusions could be reached also if other principles are applied for evaluating strategies, such as the principle of maximising expected utility, or even the principle of minimising monetary payoff.

11.4 Two-person zero-sum games

The most extensively researched kind of game is the two-person zero-sum game. It was thoroughly analysed already by von Neumann and Morgenstern in the 1940s. As explained in the taxonomy, a two-person zero-sum game is characterised by two features. First, it is always played by only two players (rather than three, or four, or n players). Second, whatever amount of utility is gained by one player is lost by the other. Hence, if the utility of an

Table 11.11

	C1	C2	C3
R1	3	−3	7
R2	4*	5	6
R3	2	7	−1

outcome for Row is 4, it is −4 for Col. This is why the game is called a zero-sum game. Many casino games are zero-sum games (even though not all of them are two-person games). For instance, if you play poker against only one opponent, you win exactly as much as your opponent loses, and vice versa.

Two-person zero-sum games are much easier to represent in decision matrices than other games. This is because if we know what the outcome for Row will be, then we automatically know the outcome for Col. Hence, two-person zero-sum games can be represented by listing only the outcomes for *one* player. By convention, we list Row's outcomes. Consider the example in Table 11.11, in which e.g. 7 in the upper rightmost corner means that Row will get 7 units of utility and Col −7 units if R1 and C3 are played.

Naturally, Row seeks to maximise the numbers in the matrix, whereas Col tries to keep the numbers as low as possible. The best outcome for Row is 7, and the worst is −3. However, the best outcome for Col is −3 (since that means +3 for her), whereas the worst is 7 (that is, −7).

Some but not all two-person zero-sum games can be solved with dominance reasoning. For example, in the game illustrated above no strategy is dominated by the others. However, despite this it is easy to figure out what rational players will do. Consider the pair (R2, C1), marked by an asterisk in the matrix. If Col *knew* that Row was going to play R2, then Col would not wish to play any other strategy, since Col tries to keep the numbers down. Furthermore, if Row *knew* that Col was going to play strategy C1, Row would have no reason to choose any other strategy. Hence, if that pair of strategies is played, no player would have any good reason to switch to another strategy, as long as the opponent sticks with his strategy. We now have to introduce a technical term.

Definition 11.1 A pair of strategies is in *equilibrium* if and only if it holds that once this pair of strategies is chosen, none of the players could reach a better outcome by *unilaterally* switching to another strategy.

Table 11.12

	C1	C2	C3		C1	C2	C3
R1	9	8	7*	R1	4	1*	2
R2	7	−5	6	R2	5	1*	9
R3	4	1	−2	R3	2	−3	−2

As explained above, the pair (R2, C1) in Table 11.11 is in equilibrium. If Col plays strategy C1, Row's best option is to play R2; and given that Row plays R2, Col could not gain anything by switching to another strategy. Neither of them has any reason to choose any other strategy, given that the other player sticks to his decision, and this is why (R2, C1) is in equilibrium.

In principle, we could solve a game by analysing each pair of pure strategies one by one and check whether it is in equilibrium. (Unfortunately, this procedure does not work for mixed strategies.) However, a much easier way to find equilibrium strategies of two-person zero-sum games is to find strategies that fulfil the *minimax condition*. This is a sufficient but not a necessary condition, i.e. it does not select all equilibria. The minimax condition tells us that *a pair of strategies are in equilibrium if (but not only if) the outcome determined by the strategies equals the minimal value of the row and the maximal value of the column.* Consider the examples in Table 11.12.

Clearly, in the leftmost example (R1, C3) is in equilibrium according to the criterion, since 7 is the minimal value of the first row, as well as the maximal value of the third column. (We indicate this by writing an asterisk in the corresponding box.) It is also fairly easy to see *why* the mimimax criterion is a sufficient criterion: Since 7 is the highest value of the column in question, Row would have no reason to choose any other strategy given that Col plays strategy C3. Furthermore, since 7 is the minimal value of its row, Col could not do better given that she knew Row was going to play strategy R1. And this is exactly why (R1, C3) is in equilibrium.

That said, in the rightmost example there are two equilibrium points: both (R1, C2) and (R2, C2) fulfil the criterion stated above. This is a potential problem, since there might be a risk that the players seek to achieve different equilibriums and consequently different strategies, and end up with an outcome that is not in equilibrium. However, it can be shown that if there are several equilibrium points, then all of them are either on the same row or in the same column. Hence, the players run no risk of failing to coordinate their strategies. For example, in the rightmost example above, it does not matter whether Row plays strategy R1 or R2; given that Col plays strategy C2 both alternatives constitute equilibrium points.

11.5 Mixed strategies and the minimax theorem

The minimax criterion stated above is not sufficient for solving every two-person zero-sum game. For example, in the game illustrated in Table 11.13 no number fulfils the criterion of being the minimal value of its row and the maximal value of its column.

The absence of a criterion for solving this game is, of course, a serious matter. Imagine that the matrix illustrates a game played by two great armies. General Row, the commander of the Horizontal army, is about to instruct his troops to attack the Vertical army by using one of the three alternative strategies R1 to R3. General Col has to defend herself against General Row's attack by choosing exactly one of the three defence strategies C1 to C3. It is evident that it is utterly important to find a solution to this game, if there is one. Indeed, the wellbeing of millions of people depends on the strategies chosen by the two generals.

Fortunately, there *is* a solution to this game. That is, there is a pair of strategies such that no player would have reason to regret their choice were that pair to be chosen. Briefly put, the optimal strategy for both generals is to roll a die! More precisely put, General Row should reason as follows: "If I throw a die and choose each pure strategy with probability 1/3, then the expected utility of my mixed strategy will be 0, no matter what probabilities General Col assigns to her strategies." Furthermore, if Row knew that Col was going to play the analogous mixed strategy, then Row would have no reason to switch to any other mixed strategy, since 0 is then the highest expected utility he could obtain, as will be explained below.

Table 11.13

	C1	C2	C3
R1	0	−100	+100
R2	+100	0	−100
R3	−100	+100	0

At this point, General Col should reason in exactly the same way, since the game is symmetric. Col reasons: "If I throw a die and choose each pure strategy with probability 1/3, then the expected utility of my choice will be 0, no matter what probabilities General Row assigns to his strategies." As before, if Col knew that Row was going to play the analogous mixed strategy, then Col would have no reason to switch to any other mixed strategy, since 0 is then the highest expected utility she would be able to obtain.

To see how General Row was able to conclude that the expected utility of playing the mixed strategy [R1 1/3, R2 1/3, R3 1/3] has an expected utility of 0, no matter what the other general decides to do, one may reason as follows. Let p denote the probability that Col chooses C1, and let q denote the probability that she chooses C2, and let r denote the probability that Col chooses C3. Then, Row's expected utility of playing [R1 1/3, R2 1/3, R3 1/3] is:

$$1/3 \left(0p - 100q + 100r\right) + 1/3 \left(100p + 0q - 100r\right) + 1/3 \left(-100p + 100q + 0r\right)$$
$$= 1/3 \left(0p - 100p + 100p - 100q + 0q + 100q + 100r - 100r + 0r\right)$$
$$= 0$$

Naturally, General Col performs the analogous calculations, which shows that the expected utility of [C1 1/3, C2 1/3, C3 1/3] is also 0.

But why is this pair of mixed strategies in equilibrium, then? Let us apply the same test as before. Given that General Col plays [C1 1/3, C2 1/3, C3 1/3], General Row *can do no better* than playing [R1 1/3, R2 1/3, R3 1/3]. For obvious mathematical reasons, no other strategy would yield a higher expected utility. Furthermore, given that General Row were to play [R1 1/3, R2 1/3, R3 1/3], Col *could not do better* by playing some other mixed or pure strategy. Hence, the pair ([R1 1/3, R2 1/3, R3 1/3], [C1 1/3, C2 1/3, C3 1/3]) is in equilibrium.

However, in this example the numbers were chosen with great care. So far we have certainly not shown that *every* two-person zero-sum game has a solution, i.e. a pair of strategies that are in equilibrium – and neither have we said anything about how one may find that solution. Fortunately, it has been established that such solutions do exist, and are not that difficult to find. More precisely put, it can be shown that every two-person zero-sum game has a solution. This means that there will always be a pair of strategies that are in equilibrium, and if there is more than one pair they all have the same expected utility. In Box 11.2 we give a formal proof of this claim.

Box 11.2   The mimimax theorem

Theorem 11.1   Every two-person zero-sum game has a solution, i.e. there is always a pair of strategies that are in equilibrium, and if there is more than one pair they all have the same expected utility.

Proof   For technical reasons, it is helpful to first consider a subset of the set of all two-person zero-sum games, viz. the set known as the *standard form* of the game. (As will be explained below, the proof can be generalised to hold for all two-person zero-sum games.) The standard form of a two-person zero-sum game has the structure shown in Table 11.14.

In this matrix, a, b, c and d are positive utilities such that $a > c$ and $b > c$ and $b > d$. It follows directly from the assumptions that no pair of *pure* strategies is in equilibrium. Therefore, if an equilibrium exists, it has to involve mixed strategies. Suppose that Row plays R1 with probability p and R2 with probability $1-p$, and that Col plays C1 with probability q and C2 with probability $1 - q$. Now, the probability that Row gets outcome a is pq, and the probability that he gets outcome d is $p \cdot (1-q)$, and so on. Hence, the expected utility for Row of playing a mixed strategy is as follows:

$$\text{Row's EU} = apq + dp(1 - q) + c(1 - p)q + b(1 - p)(1 - q) \tag{1}$$

Hence,

$$\text{Row's EU} = apq + dp - dpq + cq - cpq + b - bq - bp + bpq \tag{2}$$

$$= (a - d - c + b)pq + (d - b)p + (c - b)q + b \tag{3}$$

We now introduce a set of constants, which we define as follows:

$$K = |a - d - c + b|$$
$$L = |d - b|$$
$$M = |c - b|$$
$$N = |b|$$

Table 11.14

	C1	C2
R1	a	d
R2	c	b

Our initial assumptions that $a > c$ and $b > c$ and $b > d$ imply that $(a - d - c + b) > 0$ and $(d - b) < 0$ and $(c - b) < 0$. Hence, the formula for Row's expected utility can be rewritten as follows, where the constants K, L, M and N are the absolute values of the corresponding differences above.

$$\text{Row's EU} = Kpq - Lp - Mq + N \tag{4}$$

Note that we can rewrite this equation in the following way. (I do not expect you to see this immediately, but please feel free to check that the algebra is correct!)

$$\text{Row's EU} = K[(p - M/K)(q - L/K)] + (KN - LM)/K \tag{5}$$

Look carefully at Equation (5). For Col it is desirable to ensure that the value of (5) is as low as possible. By letting $q = L/K$, Col can ensure that the total value does not exceed $(KN - LM)/K$. Let us suppose that this indeed is part of an equilibrium strategy. Then, Row could ensure that he gets at least $(KN - LM)/K$ by letting $p = M/K$. Now, if both players play these strategies, we clearly see in equation (5) that neither of them could do better by *unilaterally* modifying their strategy. If at least one of the terms $(p - M/K)$ and $(q - L/K)$ is zero, then the value of the other term does not matter; the expected utility will nevertheless be $(KN - LM)/K$. Hence, we have shown that the following pair of mixed strategies is always an equilibrium:

$$([\text{R1}\, M/K,\ \text{R2}\, (1 - M/K)],\ [\text{C1}\, L/K,\ \text{C2}\, (1 - L/K)]) \tag{6}$$

The strategies in (6) can be used for directly finding a solution to a two-person zero-sum game. However, to complete the proof we also have to show that all two-person zero-sum games can be reduced to the standard form. This is not the right place to give a water tight proof for this claim, but some clarifying comments are in order. First, if a–d include negative utilities, we can transform the numbers into positive ones by doing a positive linear transformation of the scale. Furthermore, if the restriction that $a > c$ and $b > c$ and $b > d$ is not met we can simply switch the rows, or prove that one strategy dominates the other.   □

Exercises

11.1 Imagine that you are a fully rational decision maker facing a (one-shot) prisoner's dilemma. Explain why the outcome of the game will not be affected if you are allowed to meet and discuss with the other player before you make your move.

11.2 Can the following games be solved by applying the dominance principle? If so, what are the equilibrium strategies?

(a)

	C1	C2	C3
R1	1, 9	2, 9	0, 9
R2	3, 1	5, 2	2, 1

(b)

	C1	C2	C3
R1	1, 3	8, 2	1, 4
R2	3, 2	3, 1	2, 7
R3	1, 5	9, 2	1, 4

(c)

	C1	C2	C3
R1	1, 3	2, 4	1, 0
R2	3, 3	5, 2	0, 1
R3	2, 5	2, 0	1, 8

(d)

	C1	C2	C3	C4
R1	1, 1	1, 1	1, 4	2, 2
R2	3, 0	5, 2	2, 3	3, 3
R3	1, 1	2, 2	1, 4	8, 2
R4	0, 6	2, 4	1, 7	2, 8

11.3 Consider the suggestion that the prisoner's dilemma is a Newcomb
problem. Would an evidential decision theorist advise you to confess?
Explain!

11.4 By definition, a *non-cooperative* game is a game in which the players are
not able to form 'binding agreements'. Why can we not simply say that
a non-cooperative game is a game in which the players do not actually
cooperate?

11.5 Can the following two-person zero-sum games be solved by using the
minimax condition? If so, what are the equilibrium strategies?

(a)

	C1	C2
R1	2	−3
R2	5	7
R3	1	7

(b)

	C1	C2	C3
R1	4	−3	−7
R2	5	3	7
R3	3	2	−1

(c)

	C1	C2	C3
R1	0	−1	1
R2	1	0	−1
R3	−1	1	0

(d)

	C1	C2	C3	C4
R1	1	4	−1	1
R2	−2	3	−2	0
R3	3	6	−4	8
R4	−4	1	−3	7

11.6 The following two-person zero-sum game has no pure equilibrium strategies. Determine the mixed equilibrium strategies.

	C1	C2	C3
R1	−20	20	0
R2	20	0	−20
R3	0	−20	20

11.7 Find the mixed equilibrium strategies in the following game, where x is any positive number. (You may wish to consult the proof of the minimax theorem.)

	C1	C2
R1	3x	2x
R2	x	3x

11.8 In what sense, if any, can game theory be viewed as an instance of decision making under ignorance or risk?

11.9 Game theorists routinely assume that all players are rational and that this is common knowledge. However, sometimes players may have reason to doubt the rationality of their opponents. Suppose, for instance, that you believe the probability is seventy per cent that your opponent Col is rational and thirty per cent that she is insane, in which case she seeks to minimise rather than maximise her payoff. The payoff matrix for the game in which Col is rational looks as follows. Which strategy will be played by Row?

	C1	C2	C3
R1	0, 2	1, 1	0, 0
R2	2, 1	2, 2	1, 6
R3	0, 0	7, 1	0, 3

Solutions

11.1 The game is non-cooperative, i.e. you cannot make any binding agreements. No matter what you agree on it will always be rational to confess, since this is a dominant strategy.

11.2 (a) (R2, C2)
(b) (R2, C3)
(c) No
(d) (R2, C3)

11.3 If you decide not to confess, this piece of information is evidence for thinking that the other player will reason like you and not confess. Given that both players are fully rational, you can treat your own behaviour as evidence of what the other player will do, since the game is symmetrical.

11.4 Sometimes it is rational for rational decision makers to cooperate even when no binding agreement has been made or could have been made. For example, the crew of a sinking ship may decide to cooperate to save the ship and their own lives, without making any binding agreements.

11.5 (a) (R2, C1)

(b) (R2, C2)

(c) No

(d) (R1, C3)

11.6 ([R1 1/3, R2 1/3, R3 1/3], [C1 1/3, C2 1/3, C3 1/3])

11.7 ([R1 2/3, R2 1/3], [C1 1/3, C2 2/3])

11.8 Each player has to select a move, without knowing what the other player will do. Of course, in many decision problems the probability of the states (the move selected by the other player) is independent of what move is selected by the player. However, Newcomb-style problems indicate that this is not always the case. Interestingly enough, some authors take this line of thought to show that there is no deep, fundamental difference between game theory and the theory of individual decision making.

11.9 Row will play R2. There are two game matrices to consider, namely the one illustrated in the question and the one in which Col's utilities are reversed. If the first matrix is the true description of the game, then R2, C3 will be played. If the reversed matrix is the true matrix R2, C1 will be played.

12 Game theory II: Nonzero-sum and cooperative games

In the final sections of Chapter 11 we showed that all (two-person) zero-sum games can be solved by determining a set of equilibrium strategies. From a purely theoretical perspective, there is little more to say about zero-sum games. However, nonzero-sum games are more interesting, and require further attention by game theorists. This chapter gives a brief overview of what we currently know about nonzero-sum games. First, we shall give a proper introduction to the equilibrium concept tacitly taken for granted in the previous chapter. We will then go on to analyse a couple of well-known nonzero-sum games. The chapter ends with a discussion of whether game theory has any implications for ethics, biology and other subjects, which some scholars believe is the case.

12.1 The Nash equilibrium

The prisoner's dilemma is an example of a nonzero-sum game. This is because the loss or gain made by each player is not the exact opposite of that made by the other player. If we sum up the utilities in each box they do not always equal zero. In the prisoner's dilemma each prisoner will have to spend some time in prison no matter which strategy he chooses, so the total sum for both of them is always negative. To see this point more clearly, it is helpful to take a second look at the game matrix of the prisoner's dilemma (Table 12.1).

Obviously, R2 is dominated by R1, whereas C2 is dominated by C1. Hence, rational players will select the pair of strategies (R1, C1). It follows that (R1, C1) is in *equilibrium*, because if Row knew that Col was going to confess (play C1), then he would not be better off by switching to another strategy; and if Col knew that Row was going to confess (play R1), then he would also be best off by not switching his strategy. The notion of

Table 12.1

	C1: Confess	C2: Do not
R1: Confess	−10, −10	−1, −20
R2: Do not	−20, −1	−2, −2

equilibrium tacitly employed here was originally proposed by John Nash, a Princeton mathematician who was awarded the Nobel Prize for economics for his contributions to game theory. Nash's basic idea was very simple: Rational players will do whatever they can to ensure that they do not feel unnecessarily unhappy about their decision. Hence, if a rational player is about to play a strategy and knows that he could do something better, given that his assumptions about the opponent are held fixed, then the player will not choose the non-optimal strategy. Only strategies that fulfil this very simple 'not-better-to-switch' test will be chosen. Furthermore, this holds true for every player. Hence, rational players will always play strategies that constitute *Nash equilibriums*.

In his Ph.D. dissertation from 1950 (which is just thirty-two pages long and written in a surprisingly informal style), Nash defines the equilibrium concept as follows:

> an equilibrium point is [a set of strategies] such that each player's … strategy maximizes his pay-off if the strategies of the others are held fixed. Thus each player's strategy is optimal against those of the others. (Nash 1950a: 7)

Nash's definition is equivalent to Definition 11.1 in Section 11.4. Other game theorists have proposed alternative equilibrium concepts. However, throughout this book the term 'equilibrium' will refer to a Nash equilibrium, except for in the discussion of evolutionary game theory in Section 12.4, in which we discuss the notion of evolutionary stable strategies.

Unfortunately, many nonzero-sum games have more than one Nash equilibrium. This indicates that a game cannot be properly 'solved', i.e. we cannot figure out exactly what rational players would do, by just applying Nash's equilibrium concept. As an illustration of this, we shall return to the game of stag hunt, mentioned in Chapter 1. Recall that two members of a pre-historical society must choose to either cooperate to hunt for stag, or go and hunt for hares on their own. If the hunters cooperate and hunt for stag, each of them will get 25 kilos of meat; this is the best outcome for both

Table 12.2

	C1: Hunt stag	C2: Hunt hare
R1: Hunt stag	25, 25*	0, 5
R2: Hunt hare	5, 0	5, 5*

hunters. However, a hunter who hunts for stag when the other is hunting for hare gets nothing. This is the worst outcome. A hunter hunting for hare on his own can expect to get a hare of 5 kilos. See Table 12.2. (Exercise: Make sure you understand why this is a nonzero-sum game.)

In stag hunt there are two pure Nash equilibria, (R1, C1) and (R2, C2), marked by asterisks. If Row felt confident that Col was going to play C1, then Row would have no reason to switch from R1 to R2, and nor would Col have any reason to switch from C2 to C1 if he knew that Row was going to play R1. Hence, (R1, C1) is in equilibrium. However, note that this also holds true of the opposite corner of the matrix. If Row knew that Col was going to play C2, then Row would not be any better off were he to switch from R2 to R1, and the same is true of Col *mutatis mutandis*. Hence, if (R2, C2) is played, then each player's strategy is optimal against that of the other. (Note that in this context 'optimal' means *at least as good as*.) Hence, (R2, C2) is also a Nash equilibrium. In addition to the two pure Nash equilibria, the game also has a mixed Nash equilibrium. This is not the right place to explain how to find that equilibrium. (It will be explained in Section 12.4). For the time being, I ask the reader to trust my claim that it is (R1 1/5, C1 1/5). That is, if Col plays C1 with probability 1/5, then it would be optimal for Row to play R1 with probability 1/5, and vice versa. Each player's expected catch will then be 1/5 $(25 \cdot 1/5 + 0 \cdot 4/5) + 4/5(1/5 \cdot 5 + 4/5 \cdot 5) = 5$, and neither of them could expect to gain anything by switching to another strategy.

So what should we expect from rational agents playing stag hunt, then? Nash's equilibrium concept only gives a partial answer to this question. It eliminates those strategies that will not be played, i.e. (R1, C2) and (R2, C1), and a number of mixed strategies. But what about the three remaining equilibrium strategies? Are they all equally reasonable?

Imagine that you are Row. It then seems reasonable to look at stag hunt as a decision under risk (in which the outcome may or may not be probabilistically correlated with the strategy chosen by Col). You therefore assign a subjective probability p to the state in which Col plays C1 given that you play R1, and probability $1 - p$ to the state in which Col plays C2 given that you play R1. Furthermore, you assign a probability q to the state in which Col plays C1 given that you play R2, and a probability $1-q$ to the state in which Col plays C2 given that you play R2. If your utility of meat is linear, you will thus choose R1 over R2 if and only if $25p + 0(1-p) > 5q + 5(1-q)$. This inequality can be simplified to $p > 1/5$. Hence, if your subjective probability is higher than 1/5 that Col will hunt for stag given that you do so, then you should also hunt for stag. Furthermore, Col should reason in exactly the same way, since the game is symmetric. This gives us two clear and unambiguous recommendations, which may of course be applied to similar situations *mutatis mutandis*.

However, note that Nash's equilibrium concept played no role whatsoever in the alternative analysis. The strategy to be chosen by the other player was treated as a state of the world, to which we assigned a subjective probability. So what is the point of identifying the Nash equilibria? Why not treat (non-cooperative) nonzero-sum games as individual decisions under risk? I leave it to the reader to make up her own view about this.

That said, I would like to make the following point. In the analysis above we tacitly took for granted that rational players maximise expected utility. Some people may find that assumption questionable. A decision maker who is risk averse to a sufficiently high degree, in either of the senses explained in Section 8.5, will always hunt for hare, since that will give him 5 kilos of meat for sure, that is, he runs no risk of ending up with nothing. Therefore, each player may stop worrying about what the other player will do, since it is better to be safe than sorry! This leads us to a conclusion that is familiar from the study of the prisoner's dilemma: What is best for each (risk-averse) individual may not be best for the group as a whole. In order to reach the outcome that is best for a group playing stag hunt, we have to ensure that the players are prepared to take at least moderate risks.

Some philosophers have used stag hunt for illustrating the importance of mutual trust among fellow citizens in society. We all benefit from a society in which people cooperate, but if rational individuals are to start cooperating they have to trust each other. Furthermore, the benefits of trust always run the risk of being destroyed by too-risk-averse decision makers. If each player trusts that his opponent will be willing to cooperate and hunt for stag, then the players will be able to reach an outcome that is better for the group as a whole as well as for each and every individual. However, if there is no such mutual trust in place, then the better outcome will never be reached. Hence, it is of utter importance that we ensure that we are somehow able to establish mechanisms that preserve and increase the degree of trust we have in our fellow citizens.

Before moving on to discuss other non-cooperative games, a final remark is appropriate. Economists often talk about 'Pareto efficient' states. Briefly put, a state is Pareto efficient if and only if no one's utility level can be increased unless the utility level for someone else is decreased. (R2, C2) is not a Pareto efficient state, but (R1, C1) is. Many economists think that Pareto efficiency is a plausible minimal normative condition. Very crudely put, the idea is that all states that are not Pareto efficient should be avoided. However, as explained above, individual rationality does not by any means guarantee that society will reach a Pareto efficient state. If you believe to a sufficiently strong degree that your opponent will hunt for hare, then it is rational for you to hunt for hare too, even though the outcome will not be Pareto efficient. This is yet another illustration of the importance that society promotes trust and ensures that people are not too risk averse. Because if we do not trust that our fellow citizens will cooperate, and if we are too risk averse, then we will not reach the best outcomes.

12.2 The battle of the sexes and chicken

We shall now consider a slightly different nonzero-sum game. Unlike stag hunt, this game cannot be dealt with by 'merely' establishing a certain degree of trust among the players. Consider the following story. A newly married couple wish to spend some time together. They are thinking of either going to the opera or attending a football match. They much prefer doing something together, rather than going alone. However, Row (the wife) prefers to go to the opera rather than the football stadium, whereas Col (her husband) has the opposite preference: He prefers to watch football rather than listen to opera, but both activities are worth undertaking just in case Row comes along. For historical reasons, this game has become known as *the battle of the sexes*. However, it is interesting mainly because of

Table 12.3

	C1: Opera	C2: Football
R1: Opera	2, 1*	0, 0
R2: Football	0, 0	1, 2*

its game-theoretical structure, not because of what it tells us about the times in which it was created: the old-fashioned story about a married couple wishing to coordinate their leisure activities could be replaced with any story about two players seeking to coordinate their strategies, in which the payoff matrix looks like in Table 12.3.

The battle of the sexes is a non-cooperative game. This holds true even if we assume that the players are able to communicate and make agreements. The point is that there is no way in which players can form *binding* agreements. Of course, Row could threaten Col to treat him badly in the future if Col refuses the offer to go to the opera, but if that threat is to be taken into account the payoff matrix would be different. From a game-theoretical point of view, we would then be dealing with a different game, which is much less interesting. (To make the game look more plausible we could of course modify the story a bit. Let us suppose that the players are stuck in different parts of the city, and that they are unable to communicate because one of them forgot his or her cell phone at home. Now, this is still the same game, since its payoff matrix is exactly the same.)

The battle of the sexes has two pure equilibria, viz. (R1, C1) and (R2, C2). In addition to this, the game also has a mixed equilibrium, namely (R1 2/3, C1 1/3). So how should we expect rational players to play? Of course, each player seems to be justified in starting off from the assumption that both players are very likely to reason in the same way. However, in order to illustrate why this assumption creates problems, let us suppose that you are Row. You feel that you will choose R1, since that may give you a payoff of 2 rather than 1. Then Col will of course choose C2 for exactly the same reason. However, the problem is that then both of you will get 0, and this is bad for both of you. Furthermore, had you instead felt inclined to prefer R2, never mind why, then Col would under the current assumption prefer C1, for exactly the same reason, and this also leads to an outcome that is bad for both of you. The analogous point can also be made for a number of mixed strategies.

In the battle of the sexes, it is rather difficult to see how the players could actually manage to reach one of the Nash equilibria. Under reasonable assumptions about how the other player will reason, it seems almost certain that they will fail to reach an equilibrium point. If correct, this spells serious trouble for Nash. If he is right, rational players would never choose (R2, C1) or (R1, C2). But here we have suggested an argument for thinking that one of those pairs of strategies would actually be chosen. So what has gone wrong here? There is currently no agreement on this in the literature. I leave it to the reader to make up her own mind.

That said, let us also show how the battle of the sexes could be treated as an individual decision under risk or uncertainty. Of course, each player's choice would essentially depend on what she thinks the other player will do. For instance, if you are Row and assign probability 1 to the state in which Col plays C1 (irrespective of what you do), then you should of course play R1. Note that (R1, C1) is a Nash equilibrium, so then everything is fine. Furthermore, if you are able to assign a precise (subjective or frequentistic) probability to what you think your opponent will do, then you can calculate which mixed strategy you ought to play.

However, one can easily imagine situations in which the player is not prepared to assign any probabilities to what the opponent is likely to do. Perhaps the people who are about to meet up are not a married couple, but rather two distant relatives who have not met for years. Then it may very well make little sense to assign probabilities to the behaviour of the opponent. Put in the language of subjective probability theory, the assumption that the player's preferences over uncertain prospects are complete may be false. So what should a rational player do? Again, there is little agreement in the literature. As before, I leave it to the reader to make up her own mind.

Before closing this section we shall consider another example of a famous nonzero-sum game. This game is known as chicken, and its payoff matrix is given in Table 12.4.

Table 12.4

	C1: Straight	C2: Swerve
R1: Straight	−100, −100	1, −1*
R2: Swerve	−1, 1*	0, 0

Here is the story explaining why the game is called chicken. Two drivers are driving at high speed straight against each other. If neither of them swerves, they will both crash and die. The utility of dying is, say, -100. For both players the best outcome is achieved if and only if the other swerves, because a driver who swerves is called 'chicken'. For simplicity, we assume that the utility of being called chicken is -1, while that of not swerving if the other does is 1. If both swerve each gets 0 units of utility.

Stag hunt and the battle of the sexes are examples of *coordination* games, in which both players benefit from cooperating. Chicken is, on the other hand, an *anti-coordination* game. Both players benefit from choosing different strategies. Of course anti-coordination is also a question of coordination, so the importance of this distinction should not be exaggerated. As in the previous examples, there are two pure Nash equilibria (you should be able to find them yourselves by now) and one mixed equilibrium, viz. ([R1 1/100, R2 99/100], [C1 1/100, C2 99/100]). However, just as before, it is far from clear how rational players should reason to reach equilibrium, and which of these three points they would ultimately reach.

12.3 The bargaining problem

Up to this point we have focused on non-cooperative games. Recall that non-cooperative games are games in which the players are unable to form binding agreements on how to coordinate their actions. In this section we shall discuss a famous cooperative game, known as the bargaining game.

> Someone has put $100 on the table. Row and Col are offered to split the money between the two of them. The rules of the game are very simple. Each player has to write down his or her demand and place it in a sealed envelope. If the amounts they demand sum to more than $100 the players will get nothing. Otherwise each player will get the amount he or she demanded. The players are allowed to communicate and form whatever binding agreements they wish.

The bargaining game is an idealisation of a situation that arises frequently in many parts of society. For example, if an employer and a labour union agree on more flexible working hours the company may make a larger profit, say an additional hundred million. But how should this surplus be split between the employer and the labour union? If they are unable to

Table 12.5

	C1: $80	C2: $20
R1: $80	0, 0	$80, $20
R2: $20	$20, $80	0, 0

reach an agreement, neither of them will get anything. Or imagine two small firms that have signed a contract with a multinational corporation to supply a certain product. The project is so huge that neither of the small firms can supply the product without the help of the other. Therefore, if they do not agree to cooperate, neither of the firms will make any money. But how should they split the profit?

Before analysing the general version of the bargaining game, it is helpful to take a look at a slightly less complex version: Row and Col get the opportunity to split $100 between the two of them, but according to the rules of the simplified game, Row must either take $80 and Col $20, or Row must take $20 and Col $80. Otherwise they both end up with $0. As before, the players are allowed to communicate and bargain. The payoff matrix for the simplified game is given in Table 12.5.

The matrix illustrates two striking facts about the simplified bargaining game. First, its structure is very similar to that of some of the non-cooperative games discussed in Section 12.2. In fact, by transforming the utility scales and choosing other values in the lower right corner, the simplified game can be transformed into a game of chicken. (I leave it to the reader to verify this.) Hence, it is hardly surprising that the simplified game has more than one Nash equilibrium. The two pure ones are (R1, C2) and (R2, C1), and given that each player's utility of money is linear, the mixed solution is ([R1 8/10, R2 2/10], [C1 8/10, C2 2/10]). So it seems that we are facing the same old problem again.

Moreover, since the simplified bargaining game is just a version of a well-known non-cooperative game, it is natural to think that the distinction between non-cooperative and cooperative games is not that important after all. Many contemporary game theorists argue that cooperative games can essentially be reduced to non-cooperative ones. On their view, we cannot make sense of what it means to make a binding agreement unless we first show how such agreements can arise in non-cooperative games.

Thus, non-cooperative games are in a certain sense more fundamental than cooperative ones.

All this said about the simplified bargaining game, it is now time to analyse the general version of the problem. At first glance, things seem to be even worse here. This is because *every* possible way to split the $100 constitutes a Nash equilibrium (as long as no money is left on the table). To see this, imagine that you are Row and that you are fully certain that Col will demand $99. Then, given that Col will demand $99, your best option is to demand $1; if you decide to demand more, both would get nothing. Hence, your best response to Col's demand for $99 is to accept $1. The same holds true no matter which positive numbers $x and $100 − x > 0 we replace for $99 and $1, respectively.

So how should we solve the general bargaining problem, with its plethora of equilibria? Nash himself made a very clever proposal. He hypothesised that rational players facing the bargaining problem would agree on four axioms. The axioms stipulate some conditions for what would be required of a rational solution of the problem. Nash then went on to show that there is actually just *one* way of dividing the money that fulfils all four axioms. Hence, there is a unique best way to divide the money, and if the players are sufficiently rational they will find this solution.

If correct, Nash's solution may have important implications for ethics and political philosophy. Many philosophers believe that individuals wishing to share a common good have to agree on some substantial ethical principle, such as egalitarianism or prioritarianism. However, if Nash is right no such substantial ethical principle is necessary. As long as we agree on the *purely formal* ethical principle that any agreement made among rational individuals is morally acceptable, we will always be able to solve all problems of distributive ethics by applying Nash's axioms. Since these axioms only deal with what it is rational to do, we have reduced an ethical problem to a problem of rationality. The only ethical assumption we make is that any agreement made among rational individuals is morally acceptable. (We shall return to this assumption in the last section of the chapter.)

So what does Nash's solution to the bargaining problem look like? Briefly put, Nash showed that if Row were to gain $u(x)$ units of utility by getting $x dollars and Col $v(100 − x)$ units from getting $100 − x (where $100 − x > 0$), then rational players will always agree to divide the money such that the product $u(x) \cdot v(100 − x)$ is maximised. Furthermore, if three or more players are playing the game, they will agree to maximise the product $u(x_1) \cdot v(x_2) \cdot w(x_3) \cdot \cdots$, where the x's refer to the money received by each player. Hence, if each player's utility gain is directly proportional to the money she gets, then in a two-person case, they will get \$50 each, because $50 \cdot 50 = 2500$ is the maximal value of $x \cdot (100 - x)$ for all x such that $100 > x > 0$. This is a remarkable result. It means that there is only one *Nash solution* to the bargaining problem, despite the fact there are infinitely many *Nash equilibria*.

The axioms proposed by Nash, from which this conclusion is derived, are the following.

Nash equilibrium: Whatever rational bargainers agree on, the chosen strategies will be a Nash equilibrium.

Irrelevance of utility representation: The chosen strategies should be invariant to the choice of vNM utility function to represent a player's preferences.

Independence of irrelevant alternatives: The solution to the bargaining problem does not depend on whether irrelevant strategies (i.e. strategies that are not preferred by any player) are removed or added to the set of alternative strategies.

Symmetry: The players' strategies will be identical if and only if their utility functions are identical.

We shall refrain from stating the axioms formally, and hence also from giving a proof of the theorem. However, a few comments about the axioms might be appropriate. To start with, note that the first axiom implies that the solution the players agree on must be a Pareto optimum. Because if one player could get more utility, without any cost for the opponent, then he would rationally demand that, and the chosen strategies would not be in equilibrium. The second axiom, according to which it does not matter which interval scale the agents use for representing utility, is fairly uncontroversial. Indeed, it would be very odd to claim that the rational solution to the bargaining problem depended on how utility was represented.

The third axiom, the principle of irrelevant alternatives, is essentially the same principle we discussed in Section 3.4, in the discussion of the minimax regret rule. The basic idea is simple. Imagine that a strategy x is preferred over y, and that there is also a third alternative z that is worse than both of them. Now, if we remove strategy z from the set of alternatives, then the preference between x and y must remain the same. Nash expressed this idea in the following slightly more general way: "If the set T contains the set S and $c(T)$ is in S, then $c(T) = c(S)$", where $c(S)$ is the solution point in a set of alternative strategies S. If you found the counter example we gave to the principle of irrelevant alternatives in Chapter 3 convincing, you could easily reformulate the example (by 'reversing' the order in which the alternatives are presented) and use it against Nash's axiom.

Finally, the symmetry axiom is meant to "[express] equality of bargaining skill" (Nash 1950b: 159). To see why, imagine that two players have identical utility functions. Then, if one of them manages to get more than half of the money, the only explanation would be that that player's bargaining skills are better. However, in an idealised analysis like the present one, it makes sense to assume that both players are equally clever; therefore Nash proposed that the players must have equal bargaining skills. That said, it could be objected that Nash actually assumes much more. By assuming that both players will get equally much if their utility functions are equal, Nash in effect presupposed that *all other factors than the shape of the players' utility functions are irrelevant*. Nash explicitly adopted von Neumann and Morgenstern's theory of utility. However, the vNM theory is of course fully compatible with the thought that rational players bargaining about some good ought to pay attention to other things than the utility of the good in itself. Perhaps rational players should agree to give a larger fraction to the player who *deserves* it, or to the player who *needs* it most. Of course, this does not show that Nash's assumption about equal bargaining skills is unreasonable, but the assumption is surely not as uncontroversial as it might first appear to be.

12.4 Iterated games

If you face the standard version of the prisoner's dilemma, you typically have no reason to think that you will ever face the same decision again. However, in many cases this restriction to one-shot games makes little sense. Some games are played many times. For instance, recall the example with people commuting to work in the city centre, who may choose between going by public transportation or driving their own car. They also face a version of the prisoner's dilemma, but this game is iterated everyday, week after week.

Game theorists have discovered that the equilibrium strategies will sometimes be different in iterated games. To explain why this is so, we need to distinguish between two kinds of iterated games: finitely and infinitely iterated games. As explained in Section 11.2, a finitely iterated game is a game that is iterated a finite number of times. It follows (together with the assumptions we make about the players) that as the last round of the game is played, then the players *know* that they are playing the last round of the game. However, infinitely iterated games are not like that. No matter whether the game is *actually* played an infinite number of times, there is no point in time such that the players know in advance that they are about to play the last round of the game.

The distinction between finite and infinite games makes a huge difference to the analysis of iterated games. Suppose, for example, that Row and Col play the iterated prisoner's dilemma a finite number of times. Now, as they are about to play the last round they both know that it is the last round of the game, and they both know that the other player knows this. Hence, in the last round both players will confess, that is, play the dominant strategy, since each player knows they will be better off no matter what the other player decides to do. Note that in this last round it would be pointless to worry about how the opponent may react to a move in the next round, since there is no next round. Furthermore, as the players are about to play the penultimate round, they both know that each of them will confess in the last round. Hence, a player who decides to cooperate (deny the charges) knows that she will not be rewarded for this in the next round. The other player will confess no matter what happens. Knowing all this, it is clearly better for each player to confess (that is, play the dominant strategy) also in the penultimate round. By using backwards induction, it follows that both players will confess also in the pre-penultimate round of the game, and so on and so forth. The conclusion of this analysis is simple: In a finitely iterated prisoner's dilemma rational players will behave in exactly the same way as in the one-shot version of the game. Each player is best off confessing, because that will yield the best outcome for the individual. However, from a group perspective the players are of course much worse off. It would be better for both of them if they could somehow cooperate. This is the same old clash between individual rationality and group rationality we encountered in the analysis of the one-shot prisoner's dilemma.

Now consider the infinitely iterated prisoner's dilemma. As pointed out above, we need not stipulate that the game will *actually* be iterated an infinite number of times. The crucial assumption is that there is no point at which the players know in advance that the next round of the game will be the last one. Every time the players make their moves, they believe the game will be played again, against the same opponent. This stops the backwards induction argument from getting off the ground. In an infinitely iterated game each player may instead adjust his next move to what the opponent did in the previous round. In a certain sense, the players may observe each other's behaviour and learn more about what strategies are most likely to be played by the opponent.

Here is an example. If my opponent in the infinitely iterated prisoner's dilemma has been cooperative and played the mutually beneficial strategy ninety-nine times out of a hundred in the past, I may conclude that she is likely to cooperate next time too. But does this mean I should cooperate? Of course, in a one-shot prisoner's dilemma it is always better for me not to cooperate (by confessing), no matter what I think my opponent will do. However, since I am now playing an infinitely iterated prisoner's dilemma, I must also take into account how my opponent will 'punish' me in the next round if I confess in this round. Suppose that the opponent has been cooperative in the past *just because* I have also been cooperative in the past. Then it is probably not worth trying to gain a small amount of utility in this round by playing the myopically dominant strategy. An effect of that may be that the opponent stops cooperating in the next round. So it is better for me to cooperate, even though I would be better off in this round by not cooperating. Game theorists have proposed a clever strategy that explicitly takes this kind of reasoning into account, called *tit for tat*. It is defined as follows: always cooperate in the first round, and thereafter adjust your behaviour to whatever your opponent did in the previous round.

Thus, in the second round you should cooperate if and only if the opponent cooperated in the first round, and refrain from doing so ('confess') if the opponent did so in the previous round. If both players adopt this strategy they will achieve outcomes that are much better than those achieved by players always playing the dominant, non-cooperative strategy. Furthermore, the tit-for-tat strategy can be modified in various ways, by introducing probabilistic conditions. For example, in real life it would make sense to introduce a condition that prevents a player from always playing a non-cooperative strategy if the opponent failed to cooperate. Perhaps the opponent accidentally made a mistake, without having any intention of not trying to cooperate. (This is sometimes called the 'trembling hand' hypothesis.) Hence, even if the opponent did not cooperate and you are playing tit for tat, it might be wise to modify your strategy such that the probability that you cooperate against a non-cooperative player is not equal to zero. Perhaps you should give him a ten per cent chance to show that he actually prefers to cooperate and that his previous move was caused by an unintentional mistake.

Does all this mean that it is *rational* to play tit for tat in an infinitely iterated prisoner's dilemma? Note that if both players were to play tit for tat they would actually reach a Nash equilibrium. That is, for each player it holds true that if he thinks that the opponent will play tit for tat, then he is also best off by playing tit for tat. Unfortunately, this is not a convincing argument for claiming that tit for tat is rational, since there are many other Nash equilibria. For example, if you knew that your opponent was always going to play the non-cooperative move, then you would be best off by also playing the non-cooperative move. Therefore, the pair of strategies in which both players choose non-cooperative moves (by always 'confessing') also constitutes a Nash equilibrium.

In fact, the situation is even worse. According to the *folk theorems* of game theory – so called because they formalise pieces of knowledge known by game theorists long before the theorems were formally proved and published – it holds that virtually any 'feasible' payoff can be an equilibrium outcome, given that the player is sufficiently 'patient'. This is not the right place to explain in detail what the technical terms 'feasible' and 'patient' mean, since that requires more advanced mathematics than is appropriate for this book. However, the overall message should be clear: Iterated games have far *too many* equilibria! If we wish to provide general solutions to iterated games, we need a theory of equilibrium selection. So far no one has proposed any convincing theory for this. Empirical studies show that tit for tat does really well in game tournaments, but this is of course not sufficient reason for concluding that it is also the most *rational* strategy. There is currently a lot of research going on in this area.

12.5 Game theory and evolution

John Maynard Smith, a British biologist with a background in engineering and mathematics, proposed in the early 1970s that game theory could help shed light on evolutionary processes. Many biologists found his ideas useful, and game theory is nowadays an important tool in biology.

In evolutionary game theory, the 'players' are usually different subgroups of a species, which differ with respect to some biological characteristics. It is not necessary to assume that players (i.e. subgroups) know that they are playing a game, or that they are aware that they choose among a set of alternative strategies. The theoretical framework imported from game theory can be successfully applied without making any assumptions of *why* or *how* the players choose their strategies. However, the equilibrium concept adopted in evolutionary game theory is slightly different from Nash's equilibrium concept. It can be shown that every *evolutionary stable strategy* (ESS) is part of a Nash equilibrium, but not every Nash equilibrium has only evolutionary stable strategies. The basic idea behind the notion of evolutionary stable strategies is fairly simple. Imagine that a certain subgroup of a species, say rats living in Cambridge, can either play the hawk strategy when meeting other rats, which is a more aggressive strategy in which at least one rat is likely to be badly injured, or the dove strategy, which is more forgiving. Currently, the Cambridge rats play hawk with probability p and dove with probability $1 - p$. (This can either be taken to mean that the corresponding proportions of rats in Cambridge play these strategies, or that each individual rat randomises. In the present discussion it does not matter which interpretation we choose.) Now imagine that another group of rats arrive in Cambridge from nearby Ely. They play the hawk strategy with some other probability q and the dove strategy with probability $1 - q$. Now, the mixed strategy played by the Cambridge rats is in an *evolutionary stable equilibrium* if and only if it has a higher expected utility than that played by the Ely rats. That is, if the mixed strategy played by the Cambridge rats is in equilibrium, then it will not change after the Ely rats have arrived; on the contrary, the Ely rats will start to learn that they would fare better if they were to adopt the strategy played by the Cambridge rats. In this context, the loose term 'fare better' and the slightly more formal term 'expected utility' should, of course, be interpreted as fitness for survival.

Table 12.6

	C1: Hawk	C2: Dove
R1: Hawk	−4, −4	3, 0
R2: Dove	0, 3	2, 2

Formally, a strategy x is evolutionary stable if and only if the following conditions hold. Let EU(x,y) be the expected utility of playing strategy x against someone playing strategy y.

(i)   EU(x,x) $\geq$ EU(x,y) for all y.
(ii) EU(x,x) > EU(y,x) or EU(x,y) > EU(y,y) for all y.

In order to illustrate how Maynard Smith's equilibrium concept can be applied for explaining evolutionary processes, it is helpful to consider the hawk–dove game; see Table 12.6. A closer inspection reveals that the hawk–dove game is in fact equivalent to the game of chicken, discussed in Section 12.2. It is just the motivation of the payoff matrix that is a bit different. It goes like this: If both players choose the aggressive strategy (hawk), then they fare much worse than if they are less aggressive (i.e. play dove). However, the best outcome for each player is to play hawk while the other plays dove.

If analysed in the traditional way, the conclusion is that the hawk–dove game has three Nash equilibria, two pure and one mixed one, and that it is not clear which one rational players would choose. The two pure Nash equilibria are (R1, C2) and (R2, C1). However, before proceeding it is now time to also explain how to find the mixed Nash equilibrium. Imagine that you are an average rat from Cambridge playing against other Cambridge rats. Clearly, if the strategy played by the other rats is part of a mixed Nash equilibrium, then your expected utility of playing R1 and R2 must be equal. Otherwise you would have a decisive reason for preferring either R1 or R2, and by assumption this is not the case. Hence,

$$EU(R1) = EU(R2) \tag{1}$$

Hence, we have:

$$EU(R1) = -4q + 3 \cdot (1 - q) = 3 - 7q \tag{2}$$

$$EU(R2) = 0q + 2 \cdot (1 - q) = 2 - 2q \tag{3}$$

It follows that

$$q = 1/5 \qquad (4)$$

Hence, Rowrat can conclude that Colrat will play hawk with probability 1/5. Since the game is symmetric, it follows that Rowrat must also play hawk with probability 1/5; otherwise Colrat would have a decisive reason for playing either C1 or C2 with probability 1. Let us see how this fits with Maynard Smith's equilibrium concept. Clearly, neither of the pure Nash strategies is an evolutionary stable strategy, since they do not meet the conditions stipulated above. (I leave it to the reader to verify this.) However, playing hawk with probability 1/5 does meet the conditions, as can be easily verified by carrying out the appropriate calculations.

We are now in a position to explain what will happen if an Ely rat arrives in Cambridge. It can expect to meet a Cambridge rat playing hawk with probability 1/5. It follows from what has been said about the equilibrium concept that the best response for the Ely rat is to also play hawk with probability 1/5. Hence, if the Ely rat used to play hawk with a probability $q >$ 1/5 while living in Ely, because that was an equilibrium strategy in its old environment, then the Ely rat will on average be worse off in Cambridge than a Cambridge rat. The Ely rat is not choosing the best response. Hence, the Ely rat would do better to adjust its strategy. Otherwise it will accumulate a low average utility and ultimately go extinct in its new environment. Similarly, if the Ely rat plays hawk with some probability $q < 1/5$, it will also fare worse than the average Cambridge rat. So only those Ely rats who manage to adjust their behaviour will make it in Cambridge!

The reasoning outlined above is meant to provide a rough sketch of how tools from game theory can be applied in a biological context. Unsurprisingly, the analysis can be modified to incorporate much more sophisticated ideas, such as the thought that organisms sometimes take on different roles and play games that are asymmetric. It is beyond the scope of this brief introduction to describe how game theory can be applied to explain more complex evolutionary processes.

12.6 Game theory and ethics

Game theory is sometimes used for settling disputes within ethics and political philosophy. In this section we shall discuss to what extent, if any, such claims are warranted. Let us start with Nash's solution to the bargaining problem. Many authors have argued that this formal result provides an answer to what is perhaps the most fundamental question in distributive ethics: How should a group of people divide a set of resources among themselves? As explained above, Nash showed that rational bargainers would agree on a unique solution to the bargaining problem. All that is needed for figuring out how to rationally divide a set of resources among a group of people is information about the shape of their individual utility functions. Then, given that Nash's axioms of rationality are accepted, it follows that players will agree to divide the resources such that the product of their individual utility functions is maximised. According to some economists, this is also a strong reason for thinking that resources *ought* to be distributed in that way.

However, the professional philosopher will no doubt start to feel a bit uneasy at this point. First, it is rather tempting to criticise some of Nash's axioms, as explained above. Second, and more importantly, it seems that if the economists are right, Nash's result would constitute a counter example to the widely accepted Hume's law, which holds that no moral statement can be derived from a set of purely non-moral statements. Nash's axioms are statements about rational behaviour, not statements about what it would be morally right to do. That said, no one, not even Nash, would claim that the moral conclusion that we ought to realise from the Nash solution follows from his axioms alone. If this moral conclusion is to be entailed, we actually need to add a 'bridge-premise' that links Nash's result in game theory to ethics. In this case, the bridge-premise will hold that whatever a group of people rationally agree on ought, morally speaking, to be implemented. If this bridge-premise is added to the argument, then a moral conclusion could indeed be deduced from Nash's axioms.

So should we accept this bridge-premise closing the gap between rationality and morals? I think not. One can easily imagine normative reasons for not seeking to achieve what rational people have agreed on. For instance, what they have agreed on may have negative consequences for others. In that case every reasonable consequentialist theory of ethics should take the wellbeing of everybody into account, not only the well-being of the rational players playing the bargaining game. Furthermore, not all ethicists are as concerned with utility maximisation. Some ethicists assign moral importance to concepts such as rights, duties and virtues, which need not be considered by rational players accepting Nash's axioms. Imagine, for instance, a version of the bargaining game in which my equilibrium strategy involves ignoring duties, rights and virtues. In that case it is *rational* in the instrumental sense discussed by game theorists to play that strategy, but it would certainly not be ethical.

The prisoner's dilemma has also been the subject of much debate among moral philosophers. Of course, it is tempting to argue that players who cooperate are more ethically praiseworthy than other players. However, note that such conclusions fall outside game theory itself. Given that we accept Hume's law, no moral conclusion can follow from purely non-moral premises.

That said, game theory may nevertheless be used by ethicists for justifying ethical conclusions, in ways that are not at odds with Hume's law. A famous attempt to use tools from game theory for reasoning about ethics has been proposed by David Gauthier. Very briefly put, Gauthier argues that it is rational in the iterated prisoner's dilemma to cooperate if one thinks one is playing the game against a cooperatively minded opponent; otherwise it is rational to play non-cooperative strategies. Gauthier articulates this claim by distinguishing between two kinds of utility maximising players, viz. *straightforward maximisers* who always play the dominant strategy and refuse to cooperate, and *constrained maximisers* who cooperate with fellow constrained maximisers but not with straightforward maximisers. Having clarified this distinction, Gauthier then goes on to show that from each individual's point of view, it is rational to be a constrained maximiser rather than a restricted maximiser. Hence, it is rational to cooperate, given that all the standard assumptions of game theory are met. Furthermore, Gauthier also argues that it would be rational for each individual to cooperate, or rather adopt the normative rule to be a constrained maximiser, even if a number of the standard assumptions are not met. Those relaxations are normatively important, because it makes his results more applicable to real-world decision making.

The upshot of Gauthier's analysis is that self-interested rational constrained utility maximisers will come to agree on certain 'moral' rules. One such rule is to cooperate when facing the iterated prisoner's dilemma. Hence, in a certain sense, it is rational for every one to be moral. However, Gauthier is of course aware of Hume's law so to justify his moral conclusion he also argues for the bridge-premise, holding that whatever self-interested rational individuals agree on, this ought to be accepted as a valid moral conclusion. There is a vast scholarly literature on whether Gauthier's defence of (what I call) the bridge-premise and his general approach to ethics is successful.

Exercises

12.1 Find all pure Nash equilibria in the following games.

	C1	C2
R1	3, 5	0, 4
R2	1, 0	20, 30

	C1	C2
R1	0, 8	2, 2
R2	8, 0	3, 3

	C1	C2	C3
R1	0, 2	2, 2	2, 3
R2	4, 4	3, 3	1, 2
R3	3, 2	1, 3	0, 2

12.2 Consider the game below. It is a version of a coordination game known as 'the battle of the sexes'.
(a) What is a coordination game?
(b) Find the pure Nash equilibria of this game.

	C1	C2
R1	1, 1	4, 2
R2	2, 4	1, 1

12.3 Recall Nash's solution to the bargaining game. One of the conditions he proposed for deriving his solution was the principle of irrelevant alternatives, holding that *the solution to the bargaining problem does not depend on whether irrelevant strategies (i.e. strategies that are not preferred by any player) are removed or added to the set of alternative strategies.* Can you construct a realistic scenario in which this condition is violated?

12.4 Many games have more than one pair of equilibrium strategies. Does this show that game theory as we currently know it is an incomplete theory? Discuss.

12.5 It is sometimes pointed out that cooperative games can in a certain sense be 'reduced to', or 'derived from', non-cooperative games. What does this mean? That is, what is the relation between cooperative and non-cooperative games?

12.6 (a) What is an iterated game?
    (b) What does it mean to play tit for tat in the iterated prisoner's dilemma?

12.7 (a) What does the term 'evolutionary stable strategy' mean?
    (b) Is evolutionary game theory a descriptive theory, or a normative theory, or something else?

12.8 Find all pure and mixed Nash equilibria in the following nonzero-sum games.

(a)

	C1	C2
R1	−3, −3	3, 0
R2	0, 3	1, 1

(b)

	C1	C2
R1	0, 3	4, 2
R2	6, 2	3, 3

12.9 Does evolutionary game theory presuppose that biological entities like animals and plants make rational decisions?

12.10 Which equilibrium concept do you think makes most sense, Nash's or the ESS concept used in evolutionary game theory? Does it make sense to maintain that different equilibrium concepts should be used in different contexts? Why/why not? Discuss!

12.11 Many philosophers have attempted to establish a link between game theory and ethics. How can such attempts overcome Hume's law, according to which no moral conclusions can be derived from purely non-moral premises?

Solutions

12.1 (a) (R1, C1) and (R2, C2)

(b) (R2, C2)

(c) (R1, C3) and (R2, C1)

12.2 (a) Coordination games are games with multiple pure Nash equilibria. If played only once, we cannot foresee exactly how players will behave. This gives rise to a number of practical problems.

(b) (R1, C2) and (R2, C1)

12.3 Cf. the counter example to the analogous principle for decision making under ignorance in Chapter 2. If two players participate in a bargaining game, the relative value of some alternatives may depend on alternatives that are not chosen, simply because the relational aspects of the old alternatives may change as new alternatives are introduced.

12.4 I leave it to the reader to make up her own mind on this issue.

12.5 See Section 12.3.

12.6 (a) A game that is played several times by the same players, and is known to be played several times by the players.

(b) Always cooperate in the first round, and thereafter adjust your behaviour to whatever your opponent did in the previous round.

12.7 (a) See Section 12.5.

(b) Evolutionary game theory is descriptive.

12.8 (a) (R1, C2) and (R1 2/5, C1 2/5)

(b) (R1 1/2, C1 1/7)

12.9 No. The theory merely presupposes that biological entities behave *in accordance with* the prescriptions articulated in the theory.

12.10 I will leave it to the reader to develop their own answer.

12.11 I will leave it to the reader to develop their own answer.

13 Social choice theory

A group of friends has decided to spend their summer holiday together. However, it remains to be decided where to go. The set of alternatives includes Acapulco (a), Belize (b) and Cape Cod (c). Everyone in the group agrees that it would be fair to make the decision by voting. The group has three members, Isabelle, Joe and Klaus, and their preferences are as follows.

Isabelle prefers a to b, and b to c.	$(a \succ b \succ c)$
Joe prefers b to c, and c to a.	$(b \succ c \succ a)$
Klaus prefers c to a, and a to b.	$(c \succ a \succ b)$

It seems plausible to maintain that, *conceived of as a group*, the friends prefer a to b, because in a pairwise choice a will get two votes (from Isabelle and Klaus) whereas b will get just one (from Joe). Furthermore, the group prefers b to c, since b will get two votes (from Isabelle and Joe) and c one (from Klaus). Finally, the group also prefers c to a, since c will get two votes (from Joe and Klaus) and a one (from Isabelle). However, by now the voting rule has produced a cyclic preference ordering: a is preferred to b, and b to c, and c to a. This holds true although none of the individual preference orderings is cyclic.

The observation that the majority rule can give rise to cyclic preference orderings is not a novel discovery. This point was extensively analysed already by the French nobleman Marquis de Condorcet in the eighteenth century. Nowadays this result is known as the *voting paradox*. It serves as a classic illustration of the difficulties that arise if a group wishes to aggregate the preferences of its individual members into a joint preference ordering. Evidently, if group preferences are cyclic they cannot be choice guiding. (But why not roll a die and let chance make the decision? Answer: This does not explain *why* one option is better than another. It can hardly be better to spend the summer in Acapulco *because* a die landed in a certain way on a particular occasion.)

Social choice theory seeks to analyse collective decision problems. How should a group aggregate the preferences of its individual members into a joint preference ordering? In this context, a group could be any set of individuals, such as a married couple, a number of friends, the members of a club, the citizens of a state, or even all conscious beings in the Universe. In fact, if it is true as has been claimed by some authors, that single individuals can entertain several parallel preference orderings simultaneously, then the group could even be a single individual. ("I prefer to smoke a cigarette, but I also prefer to quit smoking.") Obviously, this suggestion requires that preferences are not defined in terms of choices, but in some other way.

Readers familiar with distributive ethics and theories such as utilitarianism, egalitarianism and prioritarianism will recognise that social choice theory is essentially addressing the same type of problem. Utilitarians argue that we should prefer one option over another if and only if its sum-total of utility exceeds that of the latter. Hence, utilitarians would invariably argue that there is a much simpler solution to the problem of social choice, viz. to just add the utilities of all individuals and pick some option that maximises the sum-total of utility. Egalitarians and prioritarians maintain that intuitions about equality should also be taken into account in the decision-making process. However, a drawback of all these proposals is that they require *interpersonal comparisons* of utility. If Anne's utility is 5 and Ben's is 6, then it makes sense to infer that the sum-total of utility is 11 only if we can somehow justify the assumption that Anne's utility is directly comparable with Ben's. For many decades, decision theorists have been quite sceptical about the possibility of making such interpersonal utility comparisons. Inspired by the logical positivists in the 1930s, it was concluded that there is no empirically meaningful way to test claims about interpersonal utility comparisons, and that there never will be one. In what follows we shall accept this dogma, even though it seems clear that *some* interpersonal comparisons are surely possible. For example, it is obvious that the utility of a transplant heart for a patient dying of a heart disease is far greater than the utility of the same transplant heart for a healthy person. Anyone who denies this suffers from an overdose of philosophical wisdom. (But the decision theorist replies: Exactly what does 'utility' mean here? What would the precise technical definition be?)

13.1 The social choice problem

Social choice theory attempts to analyse group decisions as precisely as possible. To achieve this aim, social choice theorists have to make their normative assumptions very explicit. Clearly, this is not a bad thing. By stating the normative assumptions underlying a theory as clearly as possible, it becomes easier to figure out exactly which parts of a theory one finds acceptable and which ought to be given up. Furthermore, if a theory *cannot* be stated in the precise way required by social choice theorists, then this is a strong reason for rejecting it.

A *social choice problem* is any decision problem faced by a group, in which each individual is willing to state at least ordinal preferences over outcomes. This means that each individual's preferences must satisfy the axioms stated in Section 5.1 (completeness, asymmetry and transitivity). Once all individuals have stated such ordinal preferences we have a set of *individual preference orderings*. The challenge faced by the social decision theorist is to somehow combine the individual preference ordering into a *social preference ordering*, that is, a preference ordering that reflects the preferences of the group.

A *social state* is the state of the world that includes everything that individuals care about. In the holiday example, the social states are Acapulco, Belize and Cape Cod. For another example, consider a society choosing between introducing a high tax level, moderate taxes or low taxes. In this case, each tax level corresponds to a social state. The term *social welfare function* (SWF) refers to any decision rule that aggregates a set of individual preference orderings over social states into a social preference ordering over those states. The majority rule is an example of an SWF. One may also construct more complicated voting rules, which give some individuals more votes than others, or stipulates that the group should vote repeatedly among pairwise alternatives and exclude the least popular alternative in each round.

All these concepts can, of course, be defined formally. To start with, an individual preference ordering can be conceived of as a vector (i.e. a list of ordered objects), $I = [a, b, c, d, \{e, f\}, g, h, ...]$ in which the first object is preferred to the second, and so on. If some objects are equi-preferred, then we include the set of those objects as elements of the vector, i.e. $\{e, f\}$ means that $e \sim f$, and so on. Thus, all individual preference orderings in society can be represented by a set of vectors $G = \{I, K, L, ...\}$. It is important to understand what kind of thing G is: This is a set that consists of individual preference orderings. In the holiday example, G is thus the set of the following vectors: $I = [a, b, c]$, $J = [b, c, a]$, $K = [c, a, b]$. However, instead of using this somewhat awkward notation, we will write I: $a \succ b \succ c$ and so on.

The aim of social choice theory is to analyse if and how a set of individual preference orderings G can be aggregated in a systematic manner into a social preference ordering S. From a mathematical point of view, the social preference ordering S is just yet another vector that lists the group's preference ordering over the objects its individuals hold preferences over. It is now natural to stipulate that an SWF is a function from G to S. This means that the very general question, "How should society make its decisions?" can be reformulated and rendered much more precise by asking: "Given an arbitrary set of individual preference orderings G, which SWF would produce the best social preference ordering S?"

It may be helpful to illustrate the technical concepts introduced above in an example. Imagine four individuals, M, N, O and Q, who are about to aggregate their preferences over a set of objects, $\{a, b, c, d, e, f\}$. Their individual preference orderings are as follows.

M: $a \succ b \succ c \succ d \succ e \succ f$
N: $b \succ a \succ c \succ e \succ d \succ f$
O: $a \sim b \succ d \succ c \succ e \succ f$
Q: $a \succ b \succ c \succ d \sim e \succ f$

How should we aggregate the preference orderings of M, N, O and Q into a social preference ordering S? For illustrative purposes, we shall consider the majority rule once again. We already know that this rule cannot provide a satisfactory solution to our problem, but it is helpful to gain a deeper understanding of why. So what would happen if individuals M, N, O and Q were to vote? Clearly, two of them prefer a to b, whereas one is indifferent, and one holds the opposite preference. Hence, S should list a before b, that is, society prefers a to b. By counting all the 'votes', the reader can check that a society accepting the majority rule will aggregate their preferences into the following group preference:

S: $a \succ b \succ c \succ d \succ e \succ f$

In this particular example, it turns out that the preference ordering of the group is exactly similar to that of individual M. This is of course very fortunate for M. However, note that the majority rule does not *in general* privilege individual M. For example, if individual O were to prefer *f* to *e*, then the majority would switch, and society would also start to prefer *f* to *e*. Hence, S would no longer coincide with M. From a normative point of view, this is very plausible. Social choice theorists articulate this intuition by claiming that every normatively reasonable SWF should be *non-dictatorial*, which means that it must not be the case that S *always* coincides with the preference ordering of a particular individual. For future reference we shall make this assumption explicit. Basically, what it tells us is that no individual should be allowed to be a dictator, and in order to express this in a technical language we introduce the following definition:

Definition 13.1

A group of people *D* (which may be a single-member group), which is part of the group of all individuals *G*, is *decisive with respect to* the ordered pair of social states (*a, b*) if and only if state *a* is socially preferred to *b* whenever everyone in *D* prefers *a* to *b*. A group that is decisive with respect to all pairs of social states is simply *decisive*.

Now consider the following desideratum, which is meant to capture the intuition that no individual should be allowed to be a dictator:

Non-dictatorship: No single individual (i.e. no single-member group *D*) of the group *G* is decisive.

Evidently, the majority rule meets this condition. We showed above that M is not a dictator, and it can be easily seen that no individual will be a dictator as long as the majority rule is accepted. That said, we have already outlined another reason for being a bit suspicious about the majority rule. As illustrated in the holiday example, the majority rule will sometimes generate cyclic preference orderings. Arguably, this shows that it cannot be applied as a general recipe for resolving social choice problems. Social choice theorists express this claim by explicitly stipulating that a reasonable social decision rule must produce a social preference ordering that meets the ordering conditions stated in Section 5.1. That is, for *every possible combination* of individual preference orderings, an SWF must produce a social preference ordering that is complete, asymmetric and transitive. This assumption can be formulated as follows.

Ordering: For every possible combination of individual preference
orderings, the social preference ordering must be complete,
asymmetric and transitive.

Clearly, the majority rule is ruled out by this condition, as some social
preference orderings generated by the majority rule are cyclic. However,
the modified version of the majority rule that uses some random device for
avoiding cyclic preference orderings is not ruled out. Having said that,
however, the modified majority rule is ruled out because of the fact that a
SWF is a *function* from the set of individual preference orderings to a social
preference ordering. If something is a function, it must always produce the
same output S whenever we insert the same input G, and the majority rule
does not meet this requirement.

13.2 Arrow's impossibility theorem

In his doctoral thesis, economist Kenneth Arrow (1951) proved a remarkable
theorem. It gained him instant fame, as well as the Nobel Prize for economics.
Very briefly put, what Arrow proved is the following: There is no social welfare
function (SWF) that meets the conditions of non-dictatorship and ordering, as
well as two additional conditions to be introduced shortly. This result is widely
known among academics as 'Arrow's impossibility theorem'. A natural interpretation is that social decisions can never be rationally justified, simply
because every possible mechanism for generating a social preference ordering – including the majority rule – is certain to violate at least one of Arrow's
conditions. Arrow's result received massive attention in academic circles. In
the 1960s and 1970s, many people took the theorem to prove that 'democracy
is impossible'. However, the present view is that the situation is not that bad. By
giving up or modifying some of Arrow's conditions one can formulate coherent
SWFs that are not vulnerable to impossibility results. Today, the theorem
is interesting mainly because it opened up an entirely new field of inquiry.

In order to state and prove Arrow's theorem, we must first state the
conditions it is derived from. Arrow explicitly required that if everyone in
the group prefers a to b, then the group should prefer a to b. This assumption
is known as the *Pareto* condition, after the Italian economist Vilfredo Pareto.
Below we state the Pareto condition in a somewhat unusual way, by taking
advantage of the notion of decisiveness introduced in the preceding section.

Recall that a group of people is decisive with respect to the ordered pair (a, b) if and only if it turns out that society prefers a to b whenever everyone in the group prefers a to b; and if the group is decisive with respect to all states, it is simply decisive. The Pareto condition can therefore be stated as follows.

Pareto: The group of all individuals in society is decisive.

Essentially, this formulation of the Pareto condition stipulates that if everyone in society prefers one social state over another, then society must rank those two social states in the same way. This is a weak normative assumption.

Arrow's fourth and last condition is a bit more complex. This condition is known as *independence of irrelevant alternatives*. It is different from the principle with the same name discussed in Chapter 3, which is a principle for individual decision making under ignorance. On the present version of the principle, it holds that the social ranking of, say, a and b should depend only on individual preferences over *that* pair of social states. In particular, the social ranking of a and b must *not* depend on how some third (irrelevant) social state c is ranked by the individuals. To see why this seems to be a reasonable condition, consider the following example: In the Old days, society preferred a to b, simply because Old Mark and Old Nicole agreed that a is better than b. Suppose that,

Old Mark: $a \succ b \succ c$
Old Nicole: $c \succ a \succ b$

However, in the New society things are a bit different, but the *only* difference is that object c is ranked differently:

New Mark: $c \succ a \succ b$
New Nicole: $a \succ c \succ b$

Since New Mark and New Nicole still agree that a is better than b, the New society must also prefer a to b. How c is ranked is *irrelevant* when it comes to determining the social preference between a and b. Consider the following condition.

Independence of irrelevant alternatives: If all individuals have the same preference between a and b in two different sets of individual preference orderings G and G', then society's preference between a and b must be the same in G and G'.

This condition is more controversial that the others. The problem is that it effectively excludes all SWFs that are sensitive to relational properties of the individual preference orderings. Imagine, for instance, an SWF that assigns one point to each individual's most-preferred social state, and two points to the second-best social state, and so on. The social preference ordering is then generated by adding the total number of points for each social state, and thereafter ordering them such that the most-preferred social state is the one with the lowest number of points, etc. This seems to be a fairly reasonable SWF, which tallies well with our intuitions about democracy. However, this SWF is not compatible with the condition of irrelevant alternatives. Consider the following example.

U:	$c \succ b \succ a$	U': $c \succ b \succ a$	
V:	$b \succ a \succ c$	V': $c \succ b \succ a$	
W:	$a \succ c \succ b$	W': $c \succ a \succ b$	

S:	$a \sim b \sim c$	S': $c \succ b \succ a$	

In the leftmost example, each alternative gets the same number of points. Hence, all three alternatives are equi-preferred in S. However, the rightmost list of individual preference orderings can be obtained from the leftmost ones by simply putting the c's in different positions. The order between a and b remains the same in all preference orderings to the right. Despite this, the order between a and b is not the same in S and S', which is a clear violation of independence of irrelevant alternatives. This indicates that this condition is a very strong one, and it is far from obvious that an SWF cannot be normatively reasonable unless it meets this condition. (A possible reply to this criticism is to try to weaken the condition a bit, but this is not the right place to discuss alternative formulations of the condition.)

We are now in a position to state and fully grasp Arrow's impossibility theorem. What Arrow showed was that no SWF satisfies independence of irrelevant alternatives, Pareto, non-dictatorship and the ordering condition, unless society has just one member or the number of social states is fewer than three. A proof of this claim can be found in Box 13.1.

A natural response to Arrow's theorem is to question the conditions it is derived from. If we do not believe that all conditions hold true, then the impossibility result is of little importance. As pointed out above, the most controversial condition is independence of irrelevant alternatives. Several

Box 13.1   Arrow's impossibility theorem

Suppose that there are at least three distinct social states and that the number of individuals in society is finite but greater than one. Then,

Theorem 13.1   No SWF satisfies independence of irrelevant alternatives, Pareto, non-dictatorship and the ordering condition.

The proof given below, first published by Sen, is shorter and more elegant than Arrow's original version. The main idea is to show that whenever ordering, Pareto and independence of irrelevant alternatives are satisfied, then some individual in society must be a dictator. Hence, the condition of non-dictatorship fails whenever the others are accepted. The proof is based on two simple lemmas:

> **Field-Expansion Lemma** If a group D is decisive with respect to any pair of states, then it is decisive.

Proof   Let a and b be a pair of social states such that D is decisive with respect to (a, b), and let x and y be some arbitrary social states. We now have to show that D is decisive also with respect to (x, y). First, we apply the ordering condition. Let everyone in D prefer $x \succ a \succ b \succ y$, while those not in D prefer $x \succ a$ and $b \succ y$, and rank the other pairwise combination of social states in any way whatever. Since we assumed that D is decisive with respect to (a, b), society prefers a to b. By applying Pareto, we find that society prefers x to a and b to y, because those in D and not in D agree on this. The social preference ordering is thus as follows: $x \succ a \succ b \succ y$, and because of transitivity, it follows that $x \succ y$. Now, if the social preference ordering $x \succ y$ is influenced by some *other* individual preferences than preferences over x and y (that is, preferences obtaining between other social states than x and y) then that would constitute a violation of the principle of irrelevant alternatives. Hence, by assuming that D is decisive with respect to (a, b), we have shown that society must as a consequence of this prefer $x \succ y$, given the preference orderings stipulated above. So society prefers x to y because everyone in D prefers x to y. Hence, according to the definition of decisiveness, D is decisive with respect to (x, y).   □

> **Group-Contraction Lemma** If a group D (which is not a single-person group) is decisive, then so is some smaller group contained in it.

Proof   Consider a decisive group D and partition it into two groups D' and D''. Let all individuals in D' prefer $a \succ b$ and $a \succ c$, and let all individuals in D'' prefer $a \succ b$ and $c \succ b$. There are two cases to consider. First, if $a \succ c$ in the social preference ordering, then group D' would clearly be decisive with respect to that pair, since the preferences of the other individuals played no role in the derivation of the social preference ordering. We now come to the second case. Note that group D' is *not* decisive with respect to the pair (a, b) *only if* $c \geq a$ for some non-members of D'. So let us analyse this case more closely. Clearly, $a \succ b$ in the social preference ordering, simply because D is decisive and we stipulated that $a \succ b$ in both D' and D''. From $c \geq a$ and $a \succ b$ it follows by transitivity that $c \succ b$ in the social preference ordering. However, all we assumed above was that only the members of D'' had to prefer $c \succ b$. Hence, since we were able to conclude that $c \succ b$ in the social preference ordering without making any further assumptions, group D'' is decisive with respect to that pair of states. We have now shown that in each of the two cases some subgroup of D is decisive with respect to some pair of states – if $a \succ c$ in the social preference ordering then D' is decisive, otherwise D'' is. Finally, because of the Field-Expansion Lemma, we know that a group that is decisive with respect to *some* pair of states is decisive with respect to *every* pair. Hence, if D is decisive there clearly will be some smaller subgroup of D that is also decisive.   □

Proof of Theorem 13.1.   We apply the Field-Expansion and Group-Contraction lemmas to prove the main theorem. Pareto tells us that the group of all individuals is decisive. Since the number of individuals in society was assumed to be finite we can apply it over and over again (to a smaller decisive group). At some point, we will eventually end up with a decisive single-member group, that is, a dictator.   □

people have discussed how this condition can be modified, since the impossibility result seems to get much of its strength from it. Naturally, if the independence condition is rejected, the theorem will not go through.

Another type of response to Arrow's theorem is to propose an entirely different set of conditions and argue that the new set captures our intuitions about social choice better, and then show that some SWFs do in fact satisfy the new conditions. This strategy was pursued in an early response by Kenneth May (1952), who showed that only majority rules can satisfy his set of conditions. However, to review this alternative set of conditions is beyond the scope of the present work. All in all, Arrow's theorem has been addressed in more than three thousand articles and books.

13.3 Sen on liberalism and the Pareto principle

Amartya Sen, who also won the Nobel Prize for economics for his contributions to social choice theory, has argued that the Pareto principle is incompatible with the basic ideals of liberalism. If correct, this indicates that there is either something wrong with liberalism, or the Pareto principle, or both. In this context, the term 'liberalism' refers to a very weak normative claim, which anyone who calls herself a liberal should be willing to accept. For example, consider an individual who prefers to have pink walls rather than white walls at home. In a liberal society, we should permit this somewhat unusual preference, even if the majority would prefer to see white walls. However, this absolute freedom for the individual does not apply to all choices, only to some. More precisely put, Sen proposes a minimal condition of liberalism, according to which there is at least one pair of alternatives (a, b) for each individual such that if the individual prefers a to b, then society should prefer a to b, no matter what others prefer.

Note that Sen's notion of minimal liberalism tells us nothing about what *kind* of issues individuals ought to be decisive about. The minimal condition merely guarantees that there is at least one pair of alternatives over which the individual is decisive. Of course, it need not be related to the colour of one's walls. A more plausible example might be freedom of speech: There is at least one pair of sentences such that if you prefer to proclaim a rather than b, then society should allow you to do so. However, Sen's point is that not even this minimal version of liberalism is consistent with the Pareto principle. This is of course remarkable, since the latter is regarded as relatively uncontroversial by Arrow and many other decision theorists. A possible interpretation of Sen's result is thus that we should give up our liberal ideals, or that Pareto is not as uncontroversial as we have previously thought.

We shall now state Sen's theorem more carefully. Sen proposes the following minimal condition of liberalism, which is slightly weaker than the formulation discussed above. (This formulation merely requires that two individuals, rather than everyone, are decisive with respect to at least one pair of alternatives.)

> **Minimal liberalism:** There are at least two individuals in society such that for each of them there is at least one pair of alternatives with respect to which she is decisive, that is, there is a pair *a* and *b*, such that if she prefers *a* to *b*, then society prefers *a* to *b* (and society prefers *b* to *a* if she prefers *b* to *a*).

For technical reasons, Sen also invokes a version of the ordering condition, which is slightly weaker than that used by Arrow. However, we shall refrain from stating Sen's version here, since it makes the proof of the theorem a bit more complex. In what follows, the ordering condition we are talking about is Arrow's strong version.

Now, what Sen proved is that no SWF satisfies minimal liberalism, Pareto and the ordering condition. This theorem is sometimes referred to as 'the paradox of the Paretian liberal'. Robert Nozick, one of the most well-known proponents of liberalism in recent years, comments on it in his book *Anarchy, State and Utopia*. His main point is that Sen is wrong in constructing liberalism as a property of an SWF. Instead, it is better to think of liberalism as a *constraint* on the set of alternatives that society should be allowed to make decisions about:

> Individual rights are co-possible; each person may exercise his rights as he chooses. The exercise of these rights fixes some features of the world. Within the constraints of these fixed features, a choice can be made by a social choice mechanism based upon a social ordering, if there are any choices left to make! Rights do not determine a social ordering but instead set the constraints within which a social choice is to be made, by excluding certain alternatives, fixing others, and so on. (Nozick 1974: 165–6)

Expressed in the slightly more technical vocabulary, Nozick denies a part of the ordering condition known as 'unrestricted domain'. According to Nozick, it is simply false that an SWF should be a function from all possible individual preference orderings to a social preference ordering over the same set of objects. Some of the things that individuals hold preferences about should simply not be included in the social preference ordering – decisions on those issues should be taken by the individual, not by society. Naturally, Nozick thinks that e.g. the colour of your wall belongs to the latter kind of issues. This response to Sen's theorem is no doubt very elegant. Nozick has managed to find a way to accept both conditions under attack by Sen, viz. the condition of minimal liberalism and the Pareto principle. To avoid contradiction, he instead proposes to limit the scope of the ordering condition, which has simply been taken for granted by most other scholars. From a normative point of view, this seems fairly attractive. After all, what reason do we have for thinking that a social preference ordering must include preference over everything that individuals care about?

Box 13.2   The paradox of the Paretian liberal

Theorem 13.2   No SWF satisfies minimal liberalism, Pareto and the ordering condition.

Proof   We recapitulate Sen's own proof, which is easy to follow. Let the two individuals referred to in the condition of minimal liberalism be X and Y, and let the two decisive pairs of alternatives be (a, b) and (c, d), respectively. Obviously, (a, b) and (c, d) cannot be the same pair of alternatives, because if X's preference is $a \succ b$ and Y's is $b \succ a$, then the social preference ordering would be $a \succ b \succ a$, which contradicts the ordering condition. This leaves us with two possible cases: (a, b) and (c, d) either have one element in common, or none at all. Let us first consider the case in which they have one element in common, say $a = c$. Suppose that X's preference is $a \succ b$ and that Y's is $d \succ c (= a)$. Also suppose that everyone in society, including X and Y, agrees that $b \succ d$. Because of the ordering condition, X's preference ordering is $a \succ b \succ d$, while Y's is $b \succ d \succ a$. The ordering condition guarantees that this set of individual preference orderings is included in the domain of every SWF. However, minimal liberalism entails that society prefers $a \succ b$ and $d \succ c$, and since we assumed that $c = a$, it follows that society prefers $d \succ a$. Finally, Pareto implies that society prefers $b \succ d$. Hence, the social preference ordering is $a \succ b \succ d \succ a$, which contradicts the ordering condition.

   To complete the proof, we also have to consider the case in which the pairs (a, b) and (c, d) have no common elements. Let X's preference ordering include $a \succ b$, and let Y's include $c \succ d$, and let everyone in society (including X and Y) prefer $d \succ a$ and $b \succ c$. Hence, X's preference ordering must be as follows: $d \succ a \succ b \succ c$, while Y's is $b \succ c \succ d \succ a$. However, minimal liberalism entails that society prefers $a \succ b$ and $c \succ d$, whereas Pareto entails that $d \succ a$ and $b \succ c$. It follows that the social preference ordering is $d \succ a \succ b \succ c \succ d$, which contradicts the ordering condition.   □

13.4 Harsanyi's utilitarian theorems

John C. Harsanyi, a well-known economist and fellow Nobel Prize laureate, has proposed a radically different approach to social decision making. Briefly put, he defends a utilitarian solution to the problem of social choice, according to which the social preference ordering should be entirely determined by the sum total of individual utility levels in society. For example, if a single individual strongly prefers a high tax rate over a low tax rate, and all others disagree, then society should nevertheless prefer a high tax rate given that the preference of the single individual is sufficiently strong. Here is another equally surprising utilitarian conclusion: If a doctor can save five dying patients by killing a healthy person and transplant her organs to the five dying ones – without thereby causing any negative side-effects (such as decreased confidence in the healthcare system) – then the doctor should kill the healthy person.

In order to defend his utilitarian position, Harsanyi makes a number of assumptions. First of all, he rejects Arrow's view that individual preference orderings carry nothing but ordinal information. On Harsanyi's view, it is reasonable to assume that individual preference orderings satisfy the von Neumann and Morgenstern axioms for preferences over lotteries (or some equivalent set of axioms). This directly implies that rational individuals can represent their utility of a social state on an interval scale.

> **Individual rationality:** All individual preference orderings satisfy the von Neumann and Morgenstern axioms for preferences over lotteries. (See Section 5.2.)

To render this assumption more intelligible, we may imagine hypothetical lotteries over alternative social states. Suppose, for instance, that you are offered a lottery ticket that entitles you to a fifty-fifty chance of either living in a society with a high tax rate, or in one with a low tax rate. You are then asked to compare that lottery with a 'lottery' that entitles you to live in a society with a moderate tax rate with full certainty. Which lottery would you prefer? Given that your preferences over all social states, and all lotteries over social states, satisfy von Neumann and Morgenstern's axioms (or some equivalent set of axioms) it follows that your preferences can be represented by a utility function that measures your utility on an interval scale.

The next condition proposed by Harsanyi is a bit more abstract. Briefly put, Harsanyi asks us to imagine an individual (who may or may not be a fellow citizen) who evaluates all social states from a moral point of view. Let us refer to this individual as *the Chairperson*. If the Chairperson is a fellow citizen, then he has two separate preference orderings, viz. one personal preference ordering over all states that reflects his personal preference ordering, as well as a separate preference ordering over the same set of social states that reflects the social preference ordering. It is helpful to think of the Chairperson as an individual who is chosen at random from the entire population, and who is explicitly instructed to state two parallel preference orderings, viz. a personal preference ordering and a social one. As Harsanyi puts it, the social preference ordering is the preferences the Chairperson "exhibits in those – possibly quite rare – moments when he forces a special impartial and impersonal attitude, i.e. a *moral* attitude, upon himself" (Harsanyi 1979: 293). Of course, we do not yet *know* what the social preference ordering looks like, but Harsanyi shows that we can find out surprisingly much about it, given that it fulfils a number of structural conditions. Harsanyi's research question can thus be formulated as follows: What can be concluded about the Chairperson's social preference ordering, given that it fulfils certain structural conditions?

Before answering this question, we must of course clarify the structural conditions Harsanyi impose upon the Chairperson's social preference ordering. Consider the following condition:

> **Rationality of social preferences:** The Chairperson's social preference ordering satisfies the von Neumann and Morgenstern axioms for preferences over lotteries.

In order to assess the plausibility of this condition it does not suffice to ask, "What would *I* prefer in a choice between a lottery that gives us state *a* or *b*, and a lottery that gives us *c* or *d*?" In order to assess the new condition we must rather ask, "What would *the Chairperson* prefer in a choice between a lottery that gives us state *a* or *b*, and a lottery that gives us *c* or *d*?" Of course, it might be very difficult to answer such questions. However, note that Harsanyi's theorems will go through even if we are not able to tell *what* the Chairperson would prefer. All that matters is that we somehow know that the Chairperson's preferences, whatever they are, conform to the structural conditions proposed by von Neumann and Morgenstern.

The third condition proposed by Harsanyi is the Pareto condition.

Pareto: Suppose that *a* is preferred to *b* in at least one individual preference ordering, and that there is no individual preference ordering in which *b* is preferred to *a*. Then, *a* is preferred to *b* in the Chairperson's social preference ordering. Furthermore, if all individuals are indifferent, then so is the Chairperson in his social preference ordering.

The three conditions stated above imply that the Chairperson's social preference ordering must be a weighted sum of the individual preference orderings, in which the weight assigned to each individual preference ordering represents its moral importance relative to the others. In order to show this, it is helpful to introduce a slightly more technical vocabulary. From individual rationality it follows that individual preference orderings can be represented by utility functions that measure utility on an interval scale, and from rationality of social preferences it follows that the same holds true of the social preference ordering. Let $u_i(a)$ denote individual *i*'s utility of state *a*, and let $u_s(a)$ denote the utility of *a* as reflected in the Chairperson's social preference ordering. Furthermore, let α be a real number between 0 and 1. Then,

Theorem 13.3   (Harsanyi's first theorem) Individual rationality, rationality of social preferences and Pareto together entail that:

$$u_s(a) = \sum_{i=1}^{n} \alpha_i \cdot u_i(a) \quad \text{with} \quad \alpha_i > 0 \ \text{for} \ i = 1, \dots, n \tag{1}$$

This theorem tells us that society's utility of state *a* is a weighted sum of all individuals' utility of that state. A proof will be given in Box 13.3. Meanwhile, note that the theorem does not guarantee that every individual preference ordering will be assigned the same weight. The theorem merely guarantees that each individual preference ordering is assigned *some* weight. However, utilitarians typically argue that all individual preference orderings should be assigned the *same* weight. Harsanyi thinks he can solve this problem by introducing a further assumption, which he formulates as follows.

Equal treatment of all individuals: If all individuals' utility functions $u_1, \dots,$ u_n are expressed in equal utility units (as judged by the Chairperson, based on interpersonal utility comparisons), then the Chairperson's social utility function u_c must assign the same weight to all individual utility functions.

By adding this assumption to the previous ones, the following utilitarian conclusion can be proved.

Box 13.3   Proof of Harsanyi's theorem

We shall prove both theorems simultaneously, i.e. Theorems 13.3 and 13.4, by showing that $u_s(a) = \sum_{i=1}^{n} u_i(a)$ for every social state a. Without limiting the scope of the theorems, we stipulate that all individuals use a 0 to 1 utility scale, and that the Chairperson's social utility function starts at 0. (Its upper limit may of course exceed 1.) The utility numbers assigned by the individuals to a social state can be represented by a vector, i.e. by a finite and ordered sequence of real numbers. For instance, in a society with three individuals the vector [1/3, 0, 1] represents a state in which the first individual assigns utility 1/3 to the state in question, whereas the two others assign utility 0 and 1, respectively. Let us refer to such vectors as *state vectors*, and let the term *social utility number* refer to the numerical value of the Chairperson's social utility function for a state vector. Now consider the following lemma.

> **Lemma 1** Each state vector corresponds to one and only one social utility number.

By definition, a state vector corresponds to a social state. From *rationality of social preferences* it follows that the Chairperson's social utility function assigns some number to each social state. Hence, there is at least one social utility number that corresponds to each state vector. We also need to show that there cannot exist more than one such number. Of course, two or more social states could be represented by the same state vector (if all individuals are indifferent between them), so let us suppose for *reductio* that the Chairperson's social utility function assigns different social utility numbers u and v to two different social states with the same state vector. Now, since a single state vector can represent several social states just in case every individual is indifferent between the social states, it is helpful to apply *Pareto*: Since all individuals must be indifferent between the social states represented by u and v, it follows that the Chairperson must also be indifferent; hence, u and v are the same social utility numbers.

Lemma 1 directly entails that social utility is a function of state vectors. (This is trivial: Since each state vector corresponds to exactly one social utility number, it must be possible to capture this relationship by a function.) Hence, it holds that

$$u_s(a) = f[u_1(a), \ldots, u_n(a)] \tag{1}$$

Since (1) holds for all social states a, this equation can be abbreviated as $u_s = f[u_1, \ldots, u_n]$. It remains to show that $u_s = u_1 + \cdots + u_n$. To start with, consider the following claim:

$$kf[u_1, \ldots, u_n] = f[ku_1, \ldots, ku_n], \text{where } 1 \geq k \geq 0 \tag{2}$$

Equation (2) says that it does not matter if we first multiply all individual utilities by a constant k and then apply the function f to the new state vector, or multiply k by the social utility number corresponding to the state vector. In the present exposition, Equation (2) will be accepted without proof. (For proof, see the proof of von Neumann and Morgernstern's theorem in Appendix B.) Now consider the state vector in which all individuals assign the number 0 to the state in question. Clearly, $u_s([0, \ldots, 0]) = 0$, because *Pareto* guarantees that every other state vector will be ranked above this vector, and hence assigned a number higher than 0. Next consider all unit vectors $[1, 0, \ldots, 0], [0, 1, \ldots, 0]$, and $[0, 0, \ldots, 1]$, in which exactly one ranks the state in question as the best one. Because of *equal treatment of all individuals*, u_s must assign the same social utility number to all unit vectors; let us stipulate that the number in question is 1.

In what follows, we only consider the case with a society that has two individuals. The case with societies of three or more individuals is analogous. Let u_1 and u_2 be fixed, and let L be a lottery that yields the social states $[u_1, 0]$ and $[0, u_2]$ with equal chances. From von Neumann and Morgenstern's expected utility theorem it follows that:

$$u_s(L) = (1/2)u_s([u_1, 0]) + (1/2)u_s([0, u_2]) \tag{3}$$

By applying (2) to (3) we get:

$$u_s(L) = u_s(1/2[u_1, 0] + 1/2[0, u_2]) \tag{4}$$

Note that each *individual's* expected utility of the lottery $1/2[u_1, 0] + 1/2[0, u_2]$ is $1/2u_1$ and $1/2u_2$, respectively. Hence, because of (1) it must also hold true that:

$$u_s(L) = f[(1/2)u_1, (1/2)u_2] \tag{5}$$

By applying (2) we get:

$$u_s(L) = (1/2)f[u_1, u_2] \tag{6}$$

By applying (2) again we also find that:

$$u_s([u_1, 0]) = u_1 u_s([1, 0]) = u_1 \tag{7}$$

$$u_s([0, u_2]) = u_2 u_s([0, 1]) = u_2 \tag{8}$$

By combining (7) and (8) with (3) we obtain:

$$u_s(L) = (1/2)u_1 + (1/2)u_2 \tag{9}$$

Now we are almost done. We just have to put (9) and (6) together:

$$(1/2)f[u_1, u_2] = (1/2)u_1 + (1/2)u_2 \tag{10}$$

Hence,

$$f[u_1, u_2] = u_1 + u_2 \tag{11}$$

From (11) and (1) it follows directly that $u_s = u_1 + u_2$, which completes the proof for a society with two individuals.   □

Theorem 13.4 (Harsanyi's second theorem) Given equal treatment of all individuals, the coefficients in Harsanyi's first theorem will be equal:

$$\alpha_1 = \cdots = \alpha_n \tag{2}$$

At this point it is natural to ask if Harsanyi's theorems are as powerful as they look. Has he really *proved* that society ought to distribute its resources according to utilitarian principles? Well, as shown above, the theorems do follow from the premises. Furthermore, Harsanyi's result does not violate Hume's law, according to which ethical 'ought-statements' cannot be derived from premises comprising no such ethical 'ought-statements'. The Pareto is an ethical premise, which Harsanyi uses for bridging the gap between rationality and ethics. The condition of *equal treatment of all individuals* also has some ethical content. So in one sense, Harsanyi really gives a proof of utilitarianism.

That said, no proof is better than the premises it is based upon. In Harsanyi's case, the most dubious premise is *equal treatment of all individuals*. This condition only makes sense if one believes that interpersonal comparisons of utility are possible. As pointed out above, this has been questioned by many scholars. As such, the condition does not explain how interpersonal comparisons of utility could be made; it just *presupposes* that they are somehow possible. Furthermore, one may also question the normative content of *equal treatment of all individuals*. Why should everybody be treated equally? This is a substantial ethical question, that Harsanyi (and other utilitarians) ought to argue for, and not take for granted. For example, many ethicists would surely argue that some people ought to be treated better than others (i.e. $\alpha_i > \alpha_j$ for some individuals i and j), simply because they deserve it, or have certain rights that may not be violated, or are more virtuous, etcetera. My personal view is that the condition of *equal treatment of all individuals* is far too strong. However, even if correct, Harsanyi's first theorem (which does not rely on this condition) is still interesting, because it shows that the social utility function has to be additive. This indicates that *some* consequentialist ethical theory has to be accepted by anyone who is rational, at least as long as one thinks Harsanyi's premises make sense.

Exercises

13.1 A group of four people is about to select one out of six possible social states, a, b, c, d, e or f. (a) Which state will be selected if the decision is based on a traditional voting procedure? (Does the result depend on how the voting procedure is set up, given that all people get one vote each?) (b) Which state would be selected by the maximin rule? (It prescribes that society should prefer a state in which the worst-off person is as well off as possible.)

Anne: $a \succ b \succ c \succ d \succ e$
Bert: $b \succ a \succ c \succ e \succ d$
Carl: $a \succ c \succ d \succ b \succ e$
Diana: $a \succ b \succ c \succ d \succ e$

13.2 It follows from Arrow's impossibility theorem that social decisions cannot be based on majority voting procedures. (a) Exactly what is Arrow's criticism of majority voting? (b) Do you find his criticism convincing?

13.3 The Pareto principle entails that if we can improve the situation for the richest person in society, without thereby making things worse for anyone else, we ought to do this. Suppose you object: "But if we improve the situation for the richest person that would make poor people feel even unhappier. Therefore we shouldn't do it!" Even if true, this is not a good objection to the Pareto principle. (a) Why not? (b) Suggest a relevant objection to the Pareto principle, based on the observation that it permits us to improve the situation for the richest person in society, given that this can be achieved without thereby making other people worse off.

13.4 In the proof of Arrow's theorem, what does it mean to say that a group of people D is decisive with respect to the social states (a, b)?

13.5 Sen showed that his condition of minimal liberalism is incompatible with the Pareto principle and the ordering condition. (a) Consider the following condition of *minimal socialism*. Is it compatible with the Pareto principle and the ordering condition?

   Minimal socialism: There is at least one pair of alternatives (a, b) such that no single-person group is decisive with respect to (a, b).

   (b) Is the condition of minimal socialism compatible with the condition of minimal liberalism? That is, is it possible to be both a liberal and a socialist?

13.6 According to Harsanyi, the Chairperson's social preference ordering over social states must satisfy the independence axiom. Suggest a set of social preferences that violates this axiom.

13.7 (a) Explain the link between Harsanyi's theorem and the von Neumann–Morgenstern theorem. (b) Could critics of utilitarianism accept the assumption that the Chairperson's preferences must satisfy the von Neumann–Morgenstern axioms? (c) Do you think it would be possible to replace the von Neumann–Morgenstern theorem in Harsanyi's result with some other theorem mentioned in this book?

13.8 Suppose you think all human beings have certain rights that may never be violated, such as a right not to be tortured. (a) Is this moral belief compatible with Harsanyi's axioms? (b) If not, suggest a situation in which your beliefs about the right not to be tortured comes into conflict with Harsanyi's axioms. (c) Exactly which of Harsanyi's axioms are incompatible with the belief that all human beings have a right not to be tortured?

Solutions

13.1 (a) *a* (b) *a*

13.2 (a) Majority voting does not satisfy the ordering conditions, since it can yield cyclic orderings. (b) I leave it to the reader to make up her own view about this issue.

13.3 (a) The unhappiness felt by the poor people should be included in the description of the new social state. Therefore, the poor people would actually be harmed by making the richest person better off, so the Pareto principle does not justify this. (b) One might argue that relative differences in wellbeing are normatively important, i.e. that equality matters. The Pareto principle is incompatible with many theories of equality.

13.4 See Section 13.2.

13.5 (a) No. (b) No, one can be both a liberal and a socialist.

13.6 See the discussion of Allais' paradox in Chapter 4.

13.7 (a) See Box 13.3. (b) Rawls and others who advocate the maximin view could question the continuity axiom (holding that if $A \succ B \succ C$ then there exist some p and q such that $ApC \succ B \succ AqC$). State B may simply be better than a lottery between A and some really bad outcome C, no matter how low the probability of C is. (c) The von Neumann–Morgenstern theorem can be replaced by almost any other theorem that derives a utility function from preferences, e.g. Savage's theorem. For a useful overview, see Blackorby *et al.* (1999).

13.8 a) No. (b) Imagine that the police have caught a terrorist. By torturing her, they can force her to reveal information that will save thousands of lives. Harsanyi must say this *may* be morally permissible, but if all people have a right not to be tortured it is *never* permissible to torture people, not even terrorists. (c) No individual axiom taken in isolation is incompatible with the notion of rights, although the axioms taken together are.

14 Overview of descriptive decision theory

This chapter gives an overview of how people do *actually* make decisions. The headline news is that people frequently act in ways deemed to be irrational by decision theorists. This shows that people should either behave differently, or that there is something wrong with the normative theories discussed in the preceding chapters of this book. After having reviewed the empirical findings, both conclusions will be further considered.

The interest in descriptive decision theory arose in parallel with the development of normative theories. Given the enormous influence axiomatic theories had in the academic community in the latter half of the twentieth century, it became natural to test the axioms in empirical studies. Since many decision theorists advocate (some version of) the expected utility principle, it is hardly surprising that the axioms of expected utility theory are the most researched ones. Early studies cast substantial doubt on the expected utility principle as an accurate description of how people actually choose. However, it was not until 1979 and the publication of a famous paper by Kahneman and Tversky that it finally became widely accepted that expected utility theory is a false descriptive hypothesis. Kahneman and Tversky's paper has become one of the most frequently quoted academic publications of all times. (Kahneman was awarded the Nobel Prize in economics in 2002, but Tversky died a few years earlier.) In what follows, we shall summarise their findings, as well as some later observations.

14.1 Observed violations of the expected utility principle

Kahneman and Tversky asked a group of 72 students to state a preference between prospect A and B below. It turned out that 82% of the participants preferred prospect B to A.

A: $2,500 with probability 0.33   B: $2,400 with certainty
   $2,400 with probability 0.66
   $0 with probability 0.01

It can be easily verified that the expected *monetary value* of prospect A is slightly higher than that of prospect B. However, this is of course not sufficient for concluding that the students are irrational. It cannot be excluded that they preferred B to A simply because their marginal utility is decreasing in the interval between $2,400 and $2,500. To make a stronger case against the expected utility principle, Kahneman and Tversky also asked the same group of students to state a preference between prospects C and D below.

C: $2,500 with probability 0.33   D: $2,400 with probability 0.34
    $0 with probability 0.67 $0 with probability 0.66

Now, 83% of the participants preferred C, whereas only 17% said that they preferred D. Thus, a large majority preferred B to A, but C to D. These preferences are, however, incompatible with the independence axiom, and therefore also with the expected utility principle (see Chapter 5). To understand why, note that Kahneman and Tversky's experiment is in fact a version of Allais' paradox. Consider Table 14.1.

Since the only difference between the two pairs of lotteries is the addition of 66% chance of winning $2,400 in A and B, it follows that prospect A should be preferred to B if and only if D is preferred to C, *no matter what the decision maker's utility for money is*. (For a detailed explanation of this, see Section 4.4.) This experiment has been repeated several times, with similar results, by researchers all over the world.

Kahneman and Tversky also reported that similar violations of the expected utility principle can be observed in much simpler lotteries.

Table 14.1

	Ticket 1–33	Ticket 34–99	Ticket 100
Prospect A	$2,400	$2,400	$2,400
Prospect B	$2,500	$2,400	$0
Prospect C	$2,400	$0	$2,400
Prospect D	$2,500	$0	$0

Consider the two pairs of lotteries below. For brevity, ($4,000, 0.80) is taken to represent an 80% chance of receiving $4,000.

E: ($4,000, 0.80)   F: ($3,000)
G: ($4,000, 0.20)   H: ($3,000, 0.25)

Out of a group of 95 people, 80% preferred F to E, while 65% said they preferred G to H. Again, this pattern of preference is incompatible with the expected utility principle. To see this, note that prospect G can be conceived of as 1/4 chance of obtaining prospect E, and prospect H as a 1/4 chance of obtaining prospect F. Hence, the independence axiom entails that F must be preferred to E if and only if H is preferred to G.

   Arguably, the best explanation of these violations of the expected utility principle is that in most people's view, a certain gain is worth more than an equally large expected gain. No matter what your utility of money is, to get $3,000 with certainty is worth more than getting a prospect that has the same expected utility, since a genuine prospect always involves some uncertainty. This phenomenon is known as *the certainty effect*.

   However, not all violations of the expected utility principle can be explained by the certainty effect. When confronted with the lotteries below, 86% of a group of 66 people reported that they preferred J to I, whereas 73% preferred K to L.

I: ($6,000, 0.45)   J: ($3,000, 0.9)
K: ($6,000, 0.001)   L: ($3,000, 0.002)

Note that prospect K can be conceived of as a 1/450 chance of obtaining prospect I, whereas L can be conceived of as a 1/450 chance of obtaining J. Hence, for the same reason as above, the independence axiom entails that J may be preferred to I if and only if L is preferred to K. According to Kahneman and Tversky, the best explanation of this type of violation of the expected utility principle is that most people reason incorrectly about small probabilities. The relative difference in probability between I and J and between K and L is the same. However, whereas we 'perceive' that 0.9 is twice as much as 0.45, we arguably do not 'perceive' that 0.002 is twice as much as 0.001, but we do 'perceive' that $6,000 is twice as much as $3,000. This inability to distinguish properly between small probabilities explains why many people violate the independence axiom.

Kahneman and Tversky also observed that it makes a difference whether respondents are presented with 'positive' or 'negative' prospects, even if the end states are the same. Consider the following two pairs of prospects.

In addition to whatever you own, you have been given $1,000. You are then asked to choose between: M: ($1000, 0.50) N: ($500)

In addition to whatever you own, you have been given $2,000. You are then asked to choose between: O: (−$1000, 0.50) P: (−$500)

Note that prospects M and O are equivalent. In both cases, there is a fifty-fifty chance that you at the end of the day have either $2,000 or $1,000 extra in your wallet. Furthermore, N and P are also equivalent, because in both cases it is certain that you will end up with $1,500. Despite this, when the two pairs of prospects were presented to two different groups, it turned out that 84% out of 70 people preferred N to M, whereas 69% out of 68 people preferred O to P. For the decision theorist thinking that the final outcome of a risky choice is all that matters this result is very difficult to explain away. A possible explanation of why people do not merely focus on the final outcome is that the preference for certain gains (the certainty effect) is supplemented by an equivalent aversion against certain losses. That is, O is preferred to P because people hope that they might get away with the $2,000 they already have. If they choose P it is unavoidable that they get just $1,500. This phenomenon is called the *reflection effect*.

A potential criticism of the findings summarised above is that they are based on empirical data about *hypothetical* choices. No one was actually offered real money, so perhaps the subjects did not reflect very much before they stated their preferences. However, another group of psychologists have carried out similar experiments in a casino in Las Vegas with experienced gamblers playing for real money. They also reported that an overwhelming majority of gamblers violate the expected utility principle. This indicates that it can hardly be disputed that most people frequently violate the expected utility principle.

14.2 Prospect theory

Based on their empirical observations of the reflection effect, the certainty effect and the effects of small probabilities, Kahneman and Tversky propose a descriptive theory of choice under risk called *Prospect Theory*. This theory holds that the expected utility principle should be modified by introducing two weighting functions, one for value and one for probability. More precisely put, while the expected utility of a risky prospect is $p_1 \cdot u_1 + \cdots + p_n \cdot u_n$ (where p_1 refers to the probability of the first outcome and u_1 to its utility), its *prospect value* is given by the following expression.

$$w(p_1) \cdot v(u_1) + w(p_2) \cdot v(u_2) + \cdots + w(p_n) \cdot v(u_n) \tag{1}$$

The function w in (1) is a probability weighting function. It accounts for the observation that people tend to overestimate small probabilities, but underestimate moderate and large probabilities. The exact shape of w is an empirical issue. Based on their findings, Kahneman and Tversky propose that it may look as in Figure 14.1. In the interval in which the bold line lies above the dotted line (in the lower left corner) the value of w is higher than 1. However, in the interval in which the bold line falls below the dotted line probabilities are underestimated, i.e. the value of w is below 1.

The shape of the weighting function for values, v, is more complex. Briefly put, Kahneman and Tversky argue that in all decisions under risk, individuals first determine a base-line, and thereafter evaluate outcomes either as gains or losses. For example, if you are offered a prospect in which you have a fifty-fifty chance of wining either \$1,000 or −\$1,000, the first outcome will be evaluated as a gain and the second as a loss. The value of a loss or gain is not linear, as prescribed by expected utility theory. Instead, the value of gains and losses is best represented by an S-shaped function, in which losses matter more proportionally speaking than equally large gains (Figure 14.2).

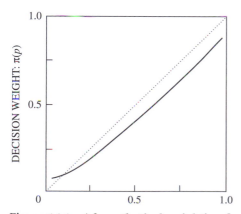

Figure 14.1   A hypothetical weighting function for probabilities.

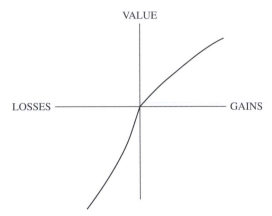

Figure 14.2   A hypothetical weighting function for values.

It should be clear that once the exact shape of the two weighting functions has been established, prospect theory can be applied to yield precise predictions of how people are likely to choose. Quite a lot of work has been done in this area, and the data obtained has led to a number of minor revisions of the theory. However, as long as we restrict the scope to the type of rather simple prospects considered here, prospects theory yields fairly accurate predictions.

14.3 Violations of transitivity and completeness

So far we have focused exclusively on the expected utility principle. However, several researchers have also reported violations of other fundamental principles of rational choice. In a classic study of transitivity, May observed that a large number of college students stated cyclic (and thus non-transitive) preferences. The subjects were asked to imagine three hypothetical marriage partners, x, y and z. They were told that x was more intelligent than y, and y more intelligent than z. They were also told that y looked better than z, and z better than x. Finally, the subjects were told that z was wealthier than x, and x wealthier than y. Out of 62 subjects, 17 stated cyclic preferences over the three candidates. A possible explanation is that the subjects may have applied the majority rule for 'voting' between each pair of attributes: Obviously, candidate x is better than y with respect to intelligence and wealth, and y is better than z with respect to intelligence and looks, whereas z is better than x with respect to looks and wealth. As you

Table 14.2

	Ticket 1–30	Ticket 31–60	Ticket 61–100
A	$54	$0	$0
B	$24	$24	$0

Table 14.3

	Ticket 1–30	Ticket 31–60	Ticket 61–100
B	$24	$24	$0
C	$4	$4	$4

may recall from Chapter 13 the majority rule sometimes yield cyclic pref-erence orderings. In several similar studies people have been asked about their preferences over more 'normal' objects. In at least one of those, it turned out that a majority of subjects stated cyclic preferences.

In a more recent study, the transitivity axiom was studied in a series of much simpler experiments. Loomes and Taylor asked 306 people taking part in a pre-university course in economics to state pairwise preferences between a series of lotteries. The subjects were asked to imagine that exactly one out of one hundred numbered tickets were to be drawn, with different payoffs for different tickets. The lotteries were represented as in Tables 14.2 and 14.3.

The meaning of the tables were explained to the subjects. For example, in lottery A one wins $54 if ticket 1 to 30 is drawn but nothing otherwise, and in lottery B one wins $24 if ticket 1 to 60 is drawn but nothing otherwise. Each subject was presented with a relatively small number of pairwise choices, including the ones illustrated in Tables 14.2 and 14.3.

It turned out that 21.6% of the 306 people completing the questionnaire stated cyclic preference orderings ($B \succ A, C \succ B, A \succ C$). In a similar study by the same authors based on a slightly different questionnaire, 18.5% stated cyclic preferences. This shows that the percentage of people with cyclic preference orderings is not negligible. Arguably, it makes little sense to claim that all those people made unintentional mistakes, or did not understand the questions. It seems to be a descriptive fact that a significant number of ordinary people fail to satisfy the transitivity axiom.

Let us now consider the completeness axiom. Do ordinary people have complete preference orderings? To begin with, it should be pointed out that it is much more difficult to test this axiom in an empirical study. This is because of the intimate relation between preference and choice taken for granted by many decision theorists: By definition, people prefer whatever they choose, and since even choosing to do nothing counts as a choice, all possible reactions to an offer to choose between, say, lotteries B and C above count as evidence of a preference. (If you simply refuse to make a choice between B and C, this just means that you are indifferent between receiving nothing, i.e. $0, and the two lotteries.) The dogma that preferences can only be elicited from choice behaviour dates back to the theory of revealed preferences proposed by the economist Paul Samuelson in the 1930s, as explained in Chapter 8. However, even if one accepts the basic empiricist spirit of the revealed preference dogma, it seems that one may nevertheless test the truth of the completeness axiom in an empirical experiment. In an influential monograph, Duncan Luce noted that people sometimes seem to choose one alternative over another with a certain probability. For example, when choosing between tea and coffee a repeated number of times, it turns out that you choose coffee over tea in $x\%$ of all choices. This need not be taken to show that your preference wobbles back and forth between coffee and tea. A more plausible explanation is that your preference is essentially probabilistic. In fact, several recent studies confirm that people do actually hold probabilistic preferences. Since the completeness axiom requires that one either prefers A to B or B to A, for all A and B, the phenomenon of probabilistic choice behaviour can be interpreted as a violation of the completeness axiom.

14.4 The relevance of descriptive decision theory

Although not all psychological data reviewed in this chapter are inconsistent with the normative prescriptions endorsed by decision theorists, some clearly are. It is simply not true that people in general follow the prescriptions derived and justified by the normative theories endorsed by decision theorists. But what follows from this observation, if anything?

Unsurprisingly, when confronted with empirical studies of how people actually do behave most decision theorists swiftly conclude that people are

Box 14.1   The psychometric paradigm

In the experiments reviewed in this chapter, subjects were asked to state preferences over lotteries with monetary outcomes. However, psychologists have also studied how people react to more complex risks, such as the risks of nuclear power, or the risk of dying in an airplane crash or in a terrorist attack. It is commonly believed that knowledge about people's reactions to various types of risks are important for guiding policy decisions.

Most studies in this area are based on questionnaires in which people are presented with a long list of risks (e.g. nuclear power meltdowns, railway accidents and terrorists attacks), and then asked to state how severe or likely they estimate each risk to be. The subjects are not given any information about the relative frequency of each risk in the past, nor about its possible consequence. The aim is to study risks as people *perceive* them, not as they really are. Unsurprisingly, the data gathered by psychologists indicate that most people tend to react to risks in ways experts find strange, or even irrational. For instance, when asked to estimate the probability of various negative events, most of us are very bad at discriminating properly between probabilities at the end points of the probability scale. Most people perceive almost no difference between a probability of one in a million and a probability of one in a billion. However, for the decision theorist wishing to compute the expected utilities there is of course a huge difference between the two. Psychological data also indicate that the perceived magnitude of a risk has little to do with its perceived acceptability. When subjects are asked to compare different risks with respect to their degree of acceptability, it turns out that the factors that matter the most to them are whether the risk is *voluntary* (nuclear power is not, but skiing is), or *natural* (nuclear power is not, but earthquakes are), or *new* (nuclear power is, but storms are not), or *catastrophic* rather than *chronic* (airplane crashes are catastrophic, but car accidents are chronic – a small number of people die in each crash, but they happen every day). None of these categories has any direct link to what decision theorists consider to be relevant, viz. the probability and utility of an outcome.

Psychologists working in this field subscribe to the so-called *psychometric paradigm*. Briefly put, the psychometric paradigm is the thought that we can explain and predict people's attitudes and reactions to risk by identifying a set of relevant factors, such as voluntariness, naturalness and novelty, and then simply measure to what degree the risk in question meets these factors. Imagine, for example, that we wish to explain why many people consider nuclear power to be too risky. If the psychometric paradigm is valid, this is because nuclear power is perceived as *unnatural*, *new*, potentially *catastrophic* and *involuntary* (on an individual level). Hence, a lobby organisation wishing to affect people's attitudes to nuclear power should therefore focus on changing our perception of the relevant factors. For example, a rational lobbyist may try to convince people that nuclear power is not a *new* technology anymore, nor is it *unnatural* since radioactive processes occur everywhere in nature. Furthermore, according to the experience we now have, the consequences of a meltdown are actually not *catastrophic*. If the lobbyist manages to change people's attitudes about these issues, the psychometric paradigm entails that they are also likely to start thinking that nuclear power is an acceptable risk.

There is some disagreement among psychologists about which factors are relevant and which are not. However, it should be emphasised that even if we were able to establish the correct list of factors, the psychometric paradigm need not be incompatible with the expected utility principle. One could simply argue that people behave *as if* they were maximising expected utility, while assigning very low utilities to certain types of outcomes. To disprove the expected utility principle one has to disprove at least one of the axioms it is derived from, and the psychometric paradigm is not in direct conflict with any of those axioms.

irrational. This conclusion of course relies heavily on the premise that our normative theories are correct, or at least cannot be falsified by empirical findings. To some extent, this seems reasonable. Nearly all philosophers accept Hume's law, which holds that no normative conclusion can be derived from purely factual premises. Thus, to determine what rational people *ought* to do, it is not sufficient to present purely factual evidence about how people do *actually* behave.

That said, one could of course maintain that empirical findings are relevant in other ways. First, knowledge about how people behave can be instrumentally valuable for politicians, policy makers, businessmen and other people who wish to adjust their courses of action to what others do. Second, knowledge about how humans make decisions are perhaps also of some intrinsic value. Third, and most importantly, some decision theorists have argued that it is meaningless to develop normative theories that we know that ordinary people *cannot* follow. 'Ought' implies 'can', as Kant famously pointed out, so if we cannot follow the prescriptions of normative decision theory it is simply false that we ought to follow them. In effect, some social scientists have developed theories of *bounded rationality* that seek to take people's limited cognitive resources (or other reasons for not being rational in the classical sense) into account in various ways. It is sometimes suggested that those theories should also be accepted as a starting point for the development of alternative normative theories.

Personally, I do not believe there is anything wrong with the logical structure of the ought-implies-can argument. If it could be convincingly shown that we cannot follow the prescriptions of our normative theories, this would surely indicate that our normative theories are flawed. Any plausible normative theory has to yield prescriptions that people can follow. However, in opposition to the proponents of bounded rationality, I do not believe that any empirical study has shown that we *cannot* obey the normative prescription proposed in the literature. The empirical studies merely show that we do not *actually* obey them. Surely, people could learn more about decision theory and eventually become more rational! In fact, this textbook is part of an attempt to make this happen.

Appendix A: Glossary

Axiom: An axiom is a fundamental premise of an argument for which no further justification is given. Example: According to the asymmetry axiom, no rational agent strictly prefers x to y and y to x.

Bargaining problem: The bargaining problem is a cooperative game with infinitely many Nash equilibria, which serves as a model for a type of situation that arises in many areas of society: A pair of players are offered to split some amount of money between the two of them. Each player has to write down his or her demand and place it in a sealed envelope. If the amounts they demand sum to more than the total amount available the players will get nothing; otherwise each player will get the amount he or she demanded. The players are allowed to communicate and form whatever binding agreements they wish. A general solution to this problem was offered by Nash, who based his proposal on a small set of intuitively plausible axioms.

Bayes' theorem: Bayes' theorem is an undisputed mathematical result about the correct way to calculate conditional probabilities. It holds that the probability of B given A equals the probability of B times the probability of A given B, divided by the following two terms: the probability of B times the probability of A given B and the probability of not-B times the probability of A given not-B. Or, put in symbols:

$$p(B|A) = \frac{p(B) \cdot p(A|B)}{[p(B) \cdot p(A|B)] + [p(\neg B) \cdot p(A|\neg B)]}$$

Bayesianism: The term 'Bayesianism' has many different meanings in decision theory, statistics and the philosophy of those disciplines. Most Bayesian accounts of decision theory and statistics can be conceived of as claims about the correct way to apply Bayes' theorem in various real-life contexts. In decision theory, Bayesianism is particularly closely associated with the view that probabilities are subjective and that rational decision makers seek to maximise subjective expected utility.

Cardinal scale: Cardinal scales are used when we measure objects numerically and differences or ratios between measurement points are preserved across all possible transformations of the scale. Example: Time, mass, money and temperature can be measured on cardinal scales. Cardinal scales can be divided into two categories, namely interval scales and ratio scales.

Decision matrix: A decision matrix is used for visualising a formal representation of a decision problem graphically. Examples of decision matrices can be found in nearly all chapters of this book. All decision matrices can be converted into decision trees (but the converse is not true).

Decision tree: A decision tree is used for visualising a formal representation of a decision problem graphically. Examples of decision trees can be found in Chapters 2, 4 and 8. All decision matrices can be converted into decision trees, but some decision trees (e.g. trees with more than one choice node) cannot be converted into decision matrices.

Dominance: Act A *strictly* dominates act B if and only if the outcome of A will be strictly better than B no matter which state of the world happens to be the true one. Act A *weakly* dominates act B if and only if the outcome of A will be as good as that of B no matter which state of the world happens to be the true one, and strictly better under at least one state.

Equilibrium: In game theory, a set of strategies is in equilibrium if and only if it holds that once these strategies are chosen, none of the players could reach a better outcome by *unilaterally* switching to another strategy. This means that each player's strategy is optimal given that the opponents stick to their chosen strategies. (See also 'Nash equilibrium'.)

Expected utility: The expected utility of an act can be calculated once we know the probabilities and utilities of its possible outcomes, by multiplying the probability and utility of each outcome and then sum all terms into a single number representing the average utility of the act.

Function (mathematical): A function is a device that takes something (such as a number) as its input and for each input returns exactly one output (such as another number). $f(x) = 3x + 7$ is a function, which returns 7 if $x = 0$ and 10 if $x = 1$, etc.

Impossibility theorem: An impossibility theorem is a formal result showing that a set of seemingly plausible premises (desiderata, axioms, etc.) imply a contradiction, and hence that no theory or principle can satisfy all premises (desiderata, axioms, etc.).

Interval scale: Interval scales measure objects numerically such that differences between measurement points are preserved across all possible transformations of the scale. Example: The Fahrenheit and Centigrade scales for measuring temperature are interval scales (but the Kelvin scale is a ratio scale). In decision theory, the most frequently used interval scale is the von Neumann–Morgenstern utility scale.

Law of large numbers: The law of large numbers is a mathematical theorem showing that if a random experiment (such as rolling a die or tossing a coin) is repeated n times and each experiment has a probability p of leading to a predetermined outcome, then the probability that the percentage of such outcomes differs from p by more than a very small amount ε converges to 0 as the number of trials n approaches infinity. This holds true for every $\varepsilon > 0$, no matter how small.

Lemma: A lemma is an intermediate result in a proof of some more complex theorem. Example: Instead of directly proving Arrow's impossibility theorem it can be broken down into several lemmas, which together entail the theorem.

Logical consequence: A conclusion B is a logical consequence of a set of premises A if and only if it can never be the case that A is true while B is false. Example: B is a logical consequence of P-and-not-P, because it can never be the case that P-and-not-P is true (so in this case the truth value of B is irrelevant).

Maximin: Maximin is a decision rule sometimes used in decisions under ignorance, which holds that one should *maximise* the *minimal* value obtainable in each decision. Hence, if the worst possible outcome of one alternative is better than that of another, then the former should be chosen.

Mixed strategy: In game theory, rational players sometimes decide what to do by tossing a coin (or by letting some other random mechanism make the decision for them). By using mixed strategies players sometimes reach equilibria that cannot be reached by playing pure strategies, i.e. non-probabilistic strategies.

Nash equilibrium: The Nash equilibrium is a central concept in game theory. A set of strategies played by a group of players constitutes a Nash equilibrium if and only if "each player's … strategy maximizes his pay-off if the strategies of the others are held fixed. Thus each player's strategy is optimal against those of the others" (Nash 1950a: 7). The definition of the term equilibrium given above is equivalent, although other definitions are also discussed in the literature.

Ordinal scale: Ordinal scales measure objects without making any comparisons of differences or ratios between measurement points across different transformations of the scale. Ordinal scales may very well be represented by numbers, but the numbers merely carry information about the relative ordering between each pair of objects. Example: I like *Carmen* more than *The Magic Flute*, and *The Magic Flute* more than *Figaro*. This ordering can be represented numerically, say by the numbers 1, 2 and 3, but the numbers do not reveal any information about how much more I like one of them over the others.

Paradox: A paradox is a false conclusion that follows logically from a set of seemingly true premises. Examples: The St Petersburg paradox, the Allais paradox, the Ellsberg paradox, and the two-envelope paradox are all well-known examples of paradoxes in decision theory.

Posterior probability: The posterior probability is the probability assigned to an event after new evidence or information about the event has been received. Example: You receive further evidence that supports your favourite scientific hypothesis. The posterior probability is the new, updated probability assigned to the hypothesis. (See also 'Bayes' theorem' and 'prior probability'.)

Preference: If you prefer A to B, i.e. if $A \succ B$, then you would choose A rather than B if offered a choice between the two. Some philosophers explain this choice-disposition by claiming that a preference is a mental disposition to choose in a certain way.

Prior probability: The prior probability is the (unconditional) probability assigned to an event before new evidence or information about the event is taken into account. Example: You receive new evidence that supports your favourite scientific hypothesis. You therefore update the probability that the hypothesis is true by using Bayes' theorem, but to do this you need to know the prior (unconditional) probability that the theory was true before the new evidence had been taken into account. (See also 'Bayes' theorem' and 'posterior probability'.)

Prisoner's dilemma: The prisoner's dilemma is a non-cooperative game in which rational players must choose strategies that they know will be sub-optimal for all players. The prisoner's dilemma illustrates a fundamental clash between individual rationality and group rationality.

Probability: The probability calculus measures how likely an event (proposition) is to occur (is to be true). Philosophers disagree about the interpretation of the probability calculus. Objectivists maintain that probabilities are objective features of the world that exist independently of us, whereas subjectivists maintain that probabilities express statements about the speaker's degree of belief in a proposition or event.

Randomised act: A randomised act is a probabilistic mixture of two or more (randomised or non-randomised) acts. Example: You have decided to do A or not-A, but instead of performing one of these acts for sure you toss a coin and perform A if and only if it lands heads up, otherwise you perform not-A.

Ratio scale: Ratio scales measure objects numerically such that ratios between measurement points are preserved across all possible transformations of the scale. Example: The widely used scales for measuring mass, length and time; in decision theory, the most frequently used ratio scale is the probability calculus, which measures how likely an event (proposition) is to occur (is to be true).

Real number: A real number is any number that can be characterised by a finite or infinite decimal representation. All rational and irrational numbers are real numbers, e.g. 2 and 4.656576786, and $\sqrt{2}$ and π. However, the imaginary numbers (e.g. $i = \sqrt{-1}$) are not real numbers.

Representation theorem: A representation theorem is a mathematical result showing that some non-numerical structure, such as preferences over a set of objects, can be represented by some mathematical structure. Example: According to the ordinal utility theorem preferences over a set of objects, $\{x, y, \ldots\}$ can be represented by a real-valued function u such that x is preferred to y if and only if $u(x) > u(y)$.

Risk: The term 'risk' has several meanings. In decisions theory, a decision under risk is taken if and only if the decision maker knows the probability and utility of all possible outcomes. In other contexts, the term 'risk' sometimes refers to the probability of an event or the expected (dis)utility of an act.

Theorem: A theorem summarises a conclusion derived from a specified set of premises. Example: The von Neumann–Morgenstern theorem holds that if a set of preferences over lotteries satisfies certain structural conditions (axioms), then these preferences can be represented by a certain mathematical structure.

Uncertainty: The term 'uncertainty' has several meanings. In decisions theory, a decision under uncertainty is taken if and only if the decision maker knows the utility of all possible outcomes, but not their probabilities. In other contexts, the term 'uncertainty' is sometimes used in a wider sense to refer to any type of situation in which there is some lack of relevant information.

Utilitarianism: Utilitarianism is the ethical theory prescribing that an act is morally right if and only if it maximises overall wellbeing. This theory was originally developed by Bentham and Mill in the nineteenth century, and is currently one of the most influential ethical theories among professional philosophers, economists and others interested in what individuals or groups of individuals ought to do and not to do.

Utility: The more you desire an object, the higher is its utility. Utility is measured on some utility scale, which is either ordinal or cardinal, and if it is cardinal it is either an interval scale or a ratio scale. (See 'cardinal scale', 'interval scale', 'ordinal scale' and 'ratio scale'.)

Zero-sum game: In a zero-sum game each player wins exactly as much as the opponent(s) lose. Most casino games and parlour games such as chess are zero-sum games, because the total amount of money or points is fixed. Games that do not fulfil this condition, such as the prisoner's dilemma, are called nonzero-sum games.

Appendix B: Proof of the von Neumann–Morgenstern theorem

The von Neumann–Morgenstern theorem (Theorem 5.2) on page 101 is an if-and-only-if claim, so we have to prove both directions of the biconditional. We first show that the axioms entail the existence part of the theorem, saying that there exists a utility function satisfying (1) and (2). In the second part of the proof, the uniqueness part, we prove that the utility function is unique in the sense articulated in (3). Finally, in the third part, we prove that if we have a utility function with properties (1)–(3) then the four axioms all hold true.

Part One

Let us start by constructing the utility function u mentioned in the theorem. Since Z is a finite set of basic prizes the completeness axiom entails that Z will contain some optimal element O and some worst element W. This means that O is preferred to or equally as good as every other element in Z, and every element in Z is preferred to or equally as good as W. Furthermore, O and W will also be the optimal and worst elements in all probabilistic mixtures of Z, i.e. in the set L. This follows from the independence axiom. To grasp why, let A be an arbitrary non-optimal element of Z such that $O \succ A$. Then, by independence, it follows that $OpC \succ ApC$, no matter what C is. Hence, $OpO \succ ApO$, and since OpO is just an alternative way of representing O, it follows that $O \succ ApO$, and this holds for every p. Hence, if we start from an optimal basic prize O, we can never construct any lottery comprising any other non-optimal basic prize A that is strictly preferred to O; this insight can be immediately generalised to hold for any combination of basic prizes.

We now stipulate that:

$$u(O) = 1 \text{ and } u(A) = 1 \text{ for every } A \sim O \tag{i}$$

$$u(W) = 0 \text{ and } u(A) = 0 \text{ for every } A \sim W \tag{ii}$$

Table B.1

	r	$p-r$	$1-p$
$0pW$	0	0	W
$0rW$	0	W	W

The next step is to define utilities for all lotteries A between 0 and W, i.e. for all lotteries that are neither optimal nor worst-case. The continuity axiom entails that for every A, there exists some p such that

$$A \sim 0pW \qquad\qquad \text{(iii)}$$

Now, if there is only one such number p we could stipulate that $u(A) = p$, for every A, $0 \succ A \succ W$. In order to see that there is in fact only one p satisfying relation (iii) for every A, suppose for *reductio* that there is some other such number $r \neq p$ such that

$$A \sim 0rW \qquad\qquad \text{(iv)}$$

From (iii) and (iv) it follows, by transitivity, that $0pW \sim 0rW$. Now, there are two cases to consider, viz. $p > r$ and $r > p$. First suppose that $p > r$. $0pW$ and $0rW$ can be represented as in Table B.1.

By applying the independence axiom from right to left, i.e. by deleting the rightmost column, we find that there exists some probability $s > 0$ (never mind what it is) such that $0 \sim 0sW$. However, we showed above that it holds for every p that $0 \succ Ap0$ given that $0 \succ A$; hence, $0 \succ 0sW$ and we have a contradiction. The analogous contradiction arises if we assume that $r > q$.

We have now shown that there exists a function u that assigns a number between 0 and 1 to every lottery, such that $u(0) = 1$, $u(W) = 0$, and for every other $A \neq 0, W$, $u(A) = p$ iff $A \sim 0pW$. The next step is to verify that the utility function we have constructed satisfies properties (1) and (2). We start with (1), i.e. the claim that

$$A \succ B \text{ if and only if } u(A) > u(B) \qquad\qquad \text{(v)}$$

We prove both directions of the biconditional simultaneously. By stipulation $u(A) = p$ iff $A \sim 0pW$, so it therefore holds that $A \sim 0u(A)W$. For the same reason, $B \sim 0u(B)W$. Hence, $A \succ B$ iff $0u(A)W \succ 0u(B)W$. By applying the independence axiom it follows that $A \succ B$ iff $u(A) > u(B)$. To see this, first suppose that $u(A) = u(B)$. It then follows that $0u(A)W \succ 0u(A)W$, and by the independence axiom we have $0 \succ 0$, which violates the asymmetry condition. Next suppose that

Table B.2

	$u(A)$	$u(B) - u(A)$	$1 - u(B)$
$Ou(A)W$	0	W	W
$Ou(B)W$	0	0	W

$u(A) < u(B)$. This is also inconsistent with the observation that $A \succ B$ iff $Ou(A)$ $W \succ Ou(B)W$. To see this, we repeat a manoeuvre that should be familiar by now. Look at Table B.2.

The headings of each column denote probabilities, not utilities. Because of the independence axiom, the rightmost column can be deleted. It then follows that there is some probability $s > 0$ (never mind what it is) such that $OsW \succ 0$. However, it was shown above that $0 \succ ApO$ for all p and A, which means that $0 \succ OsW$, and we have a contradiction.

We shall now verify property (2) of the utility function, i.e. we wish to show that

$$u(ApB) = pu(A) + (1-p)u(B) \tag{vi}$$

To start with, note that the independence axiom in conjunction with the ordering axiom guarantees that $A \sim B$ iff $ApC \sim BpC$. (Because if $A \sim B$ then neither $ApC \succ BpC$ nor $BpC \succ ApC$ can hold, so it has to be the case that $ApC \sim BpC$; the proof of the other direction of the biconditional is analogous.) It follows that

if $A \sim BpC$, then

$AqD \sim (BpC)qD$, and
$DqA \sim Dq(BpC)$ \hfill (vii)

Recall that because of the way we constructed the utility function it holds by definition that

$A \sim Ou(A)W$
$B \sim Ou(B)W$
$ApB \sim Ou(ApB)W$

Hence, by substituting $Ou(A)W$ for A and $Ou(B)W$ for B we get

$$ApB \sim Ou([Ou(A)W]p[Ou(B)W])W \tag{viii}$$

This expression might look a bit complicated, but it can fortunately be reduced to a lottery comprising only basic prizes. This is because we assumed that the probability calculus applies to lotteries, i.e. that if $pq + (1-p)r = s$, then $(AqB)p(ArB) \sim AsB$. Hence, $Ou([Ou(A)W]p[Ou(B)W])W \sim OsW$, where $s = pu(A) + (1-p)u(B)$. Because of transitivity, $ApB \sim OsW$, which entails that

$$Ou(ApB)W \sim OsW \qquad \text{(ix)}$$

Hence,

$$Ou(ApB)W \sim O[pu(A) + (1-p)u(B)]W \qquad \text{(x)}$$

By eliminating the probabilities connecting O and W we get

$$u(ApB) = pu(A) + (1-p)u(B) \qquad \text{(xi)}$$

This complets Part One of the proof.

Part Two

The aim of Part Two is to show that for every other function u' satisfying (1) and (2), there exist numbers $c > 0$ and d such that $u'(x) = c \cdot u(x) + d$. It is worth keeping in mind that u and u' are two different utility scales assigning numbers to the same set of objects, and that we already know the value of each object on the u scale. Thus, let t be a function that transforms the u scale into the u' scale. The transformation performed by t can be conceived of as a two-step process: For every number x on the u scale, t first picks a lottery A such that $u(A) = x$, i.e. $u^{-1}(x) = A$. Then, in the second step, t assigns a new real number $y = u'(A)$ to lottery A. So, by definition

$$t(x) = u'[u^{-1}(x)] = y \qquad \text{(xii)}$$

Suppose that i and j are two arbitrary numbers on the u scale such that $i \geq j$, and note that for every k, $1 \geq k \geq 0$, it holds that $i \geq ki + (1-k)j$. This means that the number $ki + (1-k)j$ is also on the u scale. Hence, by substitution

$$t(ki + (1-k)j) = u'[u^{-1}(ki + (1-k)j)] = y \qquad \text{(xiii)}$$

Now, note that $u^{-1}(ki + (1-k)j)$ is a lottery according to the definition above, and the utility of this lottery is $ki + (1-k)j$. Since i and j are also numbers on the u scale, these utilities also correspond to some lotteries, A and B, respectively, so $i = u(A)$ and $j = u(B)$. Hence,

$$t(ki + (1-k)j) = u'[u^{-1}(ku(A) + (1-k)u(B))] \qquad \text{(xiv)}$$

Because of property (2) of the theorem, it follows that

$$t(ki + (1 - k)j) = u'(AkB) \tag{xv}$$

From Part One we know that u' also satisfies (2). Hence, $t(ki + (1 - k)j) = ku'(A) + (1 - k)u'(B)$, from which it follows that

$$t(ki + (1 - k)j) = ku'(A) + (1 - k)u'(B) \tag{xvi}$$

We noted above that $i = u(A)$ and $j = u(B)$, and it follows from this that $t(i) = u'(A)$ and $t(j) = u'(B)$. By inserting this into equation (xvi), we get

$$t(ki + (1 - k)j) = kt(i) + (1 - k)t(j) \tag{xvii}$$

Now, $u'(x) = t[u(x)]$, and the right-hand side of this can be rewritten as $u'(x) = t(u(x)1 + [1 - u(x)]0)$. By applying this to equation (xvii) we get

$$u'(x) = u(x)t(1) + [1 - u(x)]t(0) = u(x)[t(1) - t(0)] + t(0) \tag{xviii}$$

Now we are almost done. We stipulate that $c = t(1) - t(0)$ and that $d = t(0)$, because we thereby get $u'(x) = cu(x) + d$. All that remains to show is that $c > 0$. This is equivalent to showing that $t(1) > t(0)$. By definition, $t(1) = u'(u^{-1}(1)) = u'(O)$, and $t(0) = u'(u^{-1}(0)) = u'(W)$. Since $O \succ W$, it follows from property (1) of the theorem that $u'(O) > u'(W)$.

Part Three

The aim of this part is to show that if we have a utility function that satisfies (1)–(3), then the four axioms hold true. Property (1) directly entails the completeness axiom. If $u(A) > u(B)$, then $A \succ B$, and if $u(B) > u(A)$ then $B \succ A$, and if $u(A) = u(B)$ then $A \sim B$. Property (1) also entails the transitivity axiom. If $u(A) > u(B)$ and $u(B) > u(C)$, then $u(A) > u(C)$. Hence, if $A \succ B$ and $B \succ C$, then $A \succ C$. To verify the independence axiom, holding that $A \succ B$ iff $ApC \succ BpC$, we need to use property (2), i.e. the fact that $u(ApB) = pu(A) + (1 - p)u(B)$. Since $u(ApC) = pu(A) + (1 - p)u(C)$ and $u(BpC) = pu(B) + (1 - p)u(C)$, all we have to show is that

$$u(A) > u(B) \text{ if and only if } pu(A) + (1 - p)u(C) > pu(B) + (1 - p)u(C) \tag{xix}$$

The truth of this biconditional can be established by observing that the term $(1 - p)u(C)$ occurs on both sides in the right-hand side inequality; hence, it can be deleted.

The fourth axiom, the continuity axiom, is derived from properties (1) and (2). If we can show that whenever $A \succ B \succ C$, there are some probabilities p and q such that $ApC \succ B \succ AqC$, we are done. We first show the implication from left to right. From (1) it follows that if $A \succ B \succ C$ then $u(A) > u(B) > u(C)$. No matter the values of $u(A)$, $u(B)$, and $u(C)$, there will exist some probabilities p and q such that

$$pu(A) + (1 - p)u(C) > u(B) > qu(A) + (1 - q)u(C)$$

Hence, because of (2), $u(ApC) > u(B) > u(AqC)$, and from (1) it follows that $ApC \succ B \succ AqC$. The proof in the other direction is analogous.

Further reading

Chapter 1

Aristotle (1958) *Topics* , trans. W.D. Ross, Oxford University Press.

Arnauld, A. and P. Nicole (1662/1996) *Logic or the Art of Thinking,* 5th edn, trans. and ed. Jill Vance Buroker, Cambridge University Press.

Hacking, I. (1975) *The Emergence of Probability: A Philosophical Study of Early Ideas about Probability, Induction, and Statistical Inference*, Cambridge University Press.

Herodotus (1954) *The Histories*, trans. Aubrey de Sélincourt, Penquin Books.

Hume, D. (1739/1888) *A Treatise of Human Nature*, ed. L.A. Selby-Bigge, Clarendon Press.

Ramsey, F.P. (1931) 'Truth and probability', in *The Foundations of Mathematics and Other Logical Essays*, Ch. VII, 156–98, ed. R.B. Braithwaite, Kegan, Paul, Trench, Trubner & Co.; Harcourt, Brace and Company.

von Neumann, J. and O. Morgenstern (1947) *Theory of Games and Economic Behavior*, 2nd edn, Princeton University Press. (1st edn without utility theory.)

Chapter 2

Bergström, L. (1966) *The Alternatives and Consequences of Actions*, Almqvist & Wiksell International.

Pascal, B. (1660/1910) *Pascal's Pensées*, trans. F.W. Trotter, Dover Classics.

Peterson, M. and S.O. Hansson (2005) 'Order-independent transformative decision rules', *Synthese* 147: 323–42.

Resnik, M. (1993) *Choices. An Introduction to Decision Theory*, University of Minnesota Press.

Savage, L.J. (1954) *The Foundations of Statistics*, John Wiley and Sons. (2nd edition 1972, Dover.)

Chapter 3

Luce, D. and H. Raiffa (1957) *Games and Decisions: Introduction and Critical Survey*, John Wiley and Sons.

Milnor, J.W. (1954) 'Games against nature', in Thrall *et al.*, *Decision Processes*, John Wiley and Sons, 49–60.

Chapter 4

Allais, M. (1953) 'Le Comportement de l'homme rationnel devant le risque: Critique des postulates et axiomes de l'ecole Américaine', *Econometrica* 21: 503–46.

Bernoulli, D. (1738/1954) 'Specimen Theoriae Novae de Mensura Sortis', *Commentari Academiae Scientiarium Imperialis Petrolitanae*, 5: 175–92. Translated as: 'Expositions of a new theory on the measurement of risk', *Econometrica* 22: 23–36.

Clark, M. and N. Schackel (2000) 'The two-envelope paradox', *Mind* 109: 415–41.

Ellsberg, D. (1961) 'Risk, ambiguity, and the Savage axioms', *Quarterly Journal of Economics* 75: 643–69.

Horgan, T. (2000) 'The two-envelope paradox, nonstandard expected utility, and the intensionality of probability', *Nous* 34: 578–603.

Jeffrey, R. (1983) *The Logic of Decision*, 2nd edn, University of Chicago Press.

Katz, B.D. and D. Olin (2007) 'A tale of two envelopes', *Mind* 116: 903–26.

Keynes, J.M. (1923) *A Treatise on Probability*, Macmillan & Co.

Levi, I. (1986) *Hard Choices: Decision Making under Unresolved Conflict*, Cambridge University Press.

Peterson, M. (2004) 'From outcomes to acts: A non-standard axiomatization of the expected utility principle', *Journal of Philosophical Logic* 33: 361–78.

Savage, L.J. (1954) *The Foundations of Statistics*, John Wiley and Sons. (2nd edition 1972, Dover.)

Chapter 5

Bentham, J. (1789/1970) *An Introduction to the Principles of Morals and Legislation*, ed. J.H. Burns and H.L.A. Hart, The Athlone Press.

Fishburn, P. (1970) *Utility Theory for Decision Making*, John Wiley and Sons. Reprinted by Krieger Press 1979.

Herstein, I.N. and J. Milnor (1953) 'An axiomatic approach to measurable utility', *Econometrica* 21: 291–7.

Krantz, D.H., R.D. Luce, P. Suppes and A. Tversky (1971) *Foundations of Measurement: Volume 1 Additive and Polynomial Representations*, Academic Press.

Levi, I. (1989) 'Rationality, prediction, and autonomous choice', *Canadian Journal of Philosophy* 19 (suppl.), 339–62. Reprinted in Levi, I. (1997) *The Covenant of Reason*, Cambridge University Press.

Luce, R.D. (1959/2005) *Individual Choice Behaviour. A Theoretical Analysis*, John Wiley and Sons.

Mill, J.S. (1863/1998) *Utilitarianism*, ed. R. Crisp, Oxford University Press.

Spohn, W. (1977) 'Where Luce and Krantz do really generalize Savage's decision model', *Erkenntnis* 11: 113–34.

von Neumann, J. and O. Morgenstern (1947) *Theory of Games and Economic Behavior*, 2nd edn, Princeton University Press. (1st edn without utility theory.)

Chapter 6

Kolmogorov, A.N. (1956) *Foundations of the Theory of Probability*, trans. N. Morrison, Chelsea Publishing Company.

Chapter 7

Carnap, R. (1950) *Logical Foundations of Probability*, University of Chicago Press.

de Finetti, B. (1931/1989) 'Probabilism: A critical essay on the theory of probability and on the value of science', *Erkenntnis* 31: 169–223. [Translation of: B. de Finetti (1931), *Probabilismo* Logos 14: 163–219.]

DeGroot, M. (1970) *Optimal Statistical Decisions*, McGraw-Hill.

Fishburn, P. (1989) 'Generalizations of expected utility theories: A survey of recent proposals', *Annals of Operations Research* 19: 3–28.

Good, I.J. (1950) *Probability and the Weighing of Evidence*, Griffin.

Hansson, B. (1975) 'The appropriateness of the expected utility model', *Erkenntnis* 9: 175–93.

Humphreys, P. (1985) 'Why propensities cannot be probabilities', *Philosophical Review* 94: 557–70.

Jeffrey, R. (1983) *The Logic of Decision*, 2nd edn (significant improvements from 1st edn), University of Chicago Press.

Keynes, J.M. (1923) *A Tract on Monetary Reform*, Macmillan & Co.

Koopman, B. (1940) 'The bases of probability', *Bulletin of the American Mathematial Society* 46: 763–74.

Kreps, D.M. (1988) *Notes on the Theory of Choice*, Westview Press.

Laplace, P.S. (1814) *A Philosophical Essay on Probabilities*, English translation 1951, Dover.

Mellor, D.H. (1971) *The Matter of Chance*, Cambridge University Press.

Popper, K. (1957) 'The propensity interpretation of the calculus of probability and the quantum theory' in *The Colston Papers 9*, ed. S. Körner, Dover, 65–70.

Ramsey, F.P. (1931) 'Truth and probability', in *The Foundations of Mathematics and Other Logical Essays*, Ch. VII, 156–98, ed. R.B. Braithwaite, Kegan, Paul, Trench, Trubner & Co.; Harcourt, Brace and Company.

Savage, L.J. (1954) *The Foundations of Statistics*, John Wiley and Sons. (2nd edn 1972, Dover.)

Schervish, M.J., T. Seidenfeld and J.B. Kadane (1990) 'State-dependent utilities', *Journal of the American Statistical Association* 85: 840–7.

Chapter 8

Arrow, K.J. (1970) 'The theory of risk aversion', in his *Essays in the Theory of Risk-Bearing*, North-Holland Publishing Company.

Blackorby, C., D. Donaldson and J.A. Weymark (1999) 'Harsanyi's social aggegation theorem for state-contingent alternatives', *Journal of Mathematical Economics* 32: 365–387.

Cubitt, R.P. (1996) 'Rational dynamic choice and expected utility theory', *Oxford Economic Papers*, New Series, 48: 1–19.

Espinoza, N. (2008) 'The small improvement argument', *Synthese* 165: 127–39.

Gärdenfors, P. and N.-E. Sahlin (1988) 'Unreliable probabilities, risk taking, and decision making', in P. Gärdenfors and N.-E. Sahlin, *Decision, Probability, and Utility*, Cambridge University Press 323–34.

Hansson, B. (1988) 'Risk aversion as a problem of conjoint measurement', in P. Gäredenfors and N.E. Sahlin, *Decision, Probability, and Utility*, Cambridge University Press.

Hansson, S.O. (2001) *The Structure of Values and Norms*, Cambridge University Press.

McClennen, E.F. (1990) *Rationality and Dynamic Choice: Foundational Explorations*, Cambridge University Press.

Pratt, J.W. (1964) 'Risk aversion in the small and in the large', *Econometrica* 32: 122–36.

Rabinowicz, W. (1995a) 'On Seinfeld's criticism of sophisticated violations of the independence axiom', *Theory and Decision* 43: 279–92.

Rabinowicz, W. (1995b) 'To have one's cake and eat it, too: Sequential choice and expected-utility violations', *Journal of Philosophy* 92: 586–620.

Samuelson, P. (1938) 'A note on the pure theory of consumer's behaviour', *Economica* 5: 61–71.

Chapter 9

Egan, A. (2007) 'Some counterexamples to causal decision theory', *Philosophical Review* 116(1): 93–114.

Gibbard, A. and W.L. Harper ([1978] 1988) 'Counterfactuals and two kinds of expected utility', in P. Gärdenfors and N.E. Sahlin, *Decision, Probability, and Utility*, Cambridge University Press, 341–76.

Harper, W. (1986) 'Mixed strategies and ratifiability in causal decision theory', *Erkenntnis* 24(1): 25–36.

Horwich, P. (1985) 'Decision theory in light of Newcomb's problem', *Philosophy of Science* 52: 431–50.

Joyce, J.M. (1999) *The Foundations of Causal Decision Theory*, Cambridge University Press.

Nozick, R. (1969) 'Newcomb's problem and two principles of choice', in *Essays in Honor of Carl G. Hempel*, ed. N. Rescher *et al.*, Reidel, 114–46.

Rabinowicz, W. (1989) 'Stable and retrievable options', *Philosophy of Science* 56: 624–41.

Skyrms, B. (1982) 'Causal decision theory', *Journal of Philosophy* 79: 695–711.

Weirich, P. (1985) 'Decision instability', *Australasian Journal of Philosophy* 63: 465–72.

Chapter 10

Blackburn, S. (1998) *Ruling Passions*, Oxford University Press.

Bradley, R. (2007) 'A unified Bayesian decision theory', *Theory and Decision* 63: 233–63.

Howson, C. and P. Urbach (2005) *Scientific Reasoning: The Bayesian Approach*, 3rd edn, Open Court Publishing Company.

Maher, P. (1993) *Betting on Theories*, Cambridge University Press.

Chapter 11

Kreps, D.M. (1990) *Game Theory and Economic Modeling*, Oxford University Press.

Lewis, D. (1979) 'Prisoner's dilemma is a Newcomb problem', *Philosophy and Public Affairs* 8: 235–40.

von Neumann, J. and O. Morgenstern (1947) *Theory of Games and Economic Behavior*, 2nd edn, Princeton University Press. (1st edn without utility theory.)

Chapter 12

Hargraves Heap, S.P. and Y. Varoufakis (1995) *Game Theory: A Critical Introduction*, Routledge.

Maynard Smith, J. (1982) *Evoloution and the Theory of Games*, Cambridge University Press.

Nash, J. (1950a). *Non-Cooperative Games*, Ph.D. dissertation Princeton University.

Nash, J. (1950b) 'The bargaining problem', *Econometrica* 18: 155–62.

Chapter 13

Arrow, K.J. (1951) *Social Choice and Individual Values*. John Wiley and Sons, New York. (2nd edn 1963.)

Broome, J. (1999) *Ethics out of Economics*, Cambridge University Press.

Condorcet (1793/1847) 'Plan de Constitution, présenté à la convention nationale les 15 et 16 février 1793', *Oeuvres* 12: 333–415.

Harsanyi, J.C. (1955) 'Cardinal welfare, individualistic ethics, and interpersonal comparisons of Utility', *Journal of Political Economy* 63: 309–21.

Harsanyi, J.C. (1979) 'Bayesian decision theory, rule utilitarianisim, and Arrow's impossibility theorem', *Theory and Decision* 11: 289–317.

May, K. (1952) 'A set of independent necessary and sufficient conditions for simple majority decision', *Econometrica* 20(4), 680–4.

Nozick, R. (1974) *Anarchy, State, Utopia,* Basic Books.

Sen, A. (1970) 'The impossibility of a Paretian liberal', *The Journal of Political Economy* 78: 152–7.

Sen, A. (1995) 'Rationality and social choice', *American Economic Review* 85: 1–24.

Sen, A. (1999) 'The possibility of social choice', *American Economic Review* 89: 349–78.

Chapter 14

Coombs, C.H., R.M. Dawes and A Tversky (1970) *Mathematical Psychology: An Elementary Introduction*, Prentice-Hall.

Kagel, J.H. and A. Roth (1995) *The Handbook of Experimental Economics*, Princeton University Press.

Kahneman, D. and A. Tversky (1979) 'Prospect theory: An analysis of decisions under risk', *Econometrica* 47: 263–91.

May, K.O. (1954) 'Intransitivity, utility, and the aggregation of preference patterns', *Econometrica* 22: 1–13.

Loomes, G. and C. Taylor (1992) 'Non-transitive preferences over gains and losses', Economic Journal 102: 357–65.

Slovic, P. (2000) *The Perception of Risk*, Earthscan.

Index